Golden Ghetto

Golden Ghetto

How the Americans & French Fell In & Out of Love
During the Cold War

STEVE BASSETT

XENO

Book design and layout by Skyler Schulze

ISBN 978-1-939096-24-1 (tradepaper)
ISBN 978-1-939096-36-4 (clothbound)
Library of Congress Cataloging-in-Publication Data
Bassett, Steve.
  Golden ghetto : how the Americans & French fell in & out of love during the Cold
War / Steve Bassett.—First Edition.
    pages cm
  Includes bibliographical references and index.
  1. Châteauroux Air Station (Châteauroux, France)—History. 2. Americans—France—
Châteauroux Region—History—20th century. 3. United States. Air Force—History—
20th century. 4. Air bases—France—History—20th century. 5. Châteauroux Region
(France)—History—20th century. 6. United States—Relations—France. 7. France—
Relations—United States. I. Title.
  UG635.F82C5225 2013
  358.4'170944551—dc23
                        2013005929
The National Endowment for the Arts, the Los Angeles County Arts Commission, the
Los Angeles Department of Cultural Affairs, the City of Pasadena Cultural Affairs
Division, Sony Pictures Entertainment, and the Dwight Stuart Youth Fund partially
support Red Hen Press.

First Edition
XENO Books is an imprint of Red Hen Press, Pasadena, CA
www.redhen.org/xeno

ACKNOWLEDGMENTS

This book would not have been possible without the commitment and inspiration supplied by my wife, Darlene Chandler Bassett, who for seven years demanded my very best, no shortcuts or excuses accepted. Christine Cappuccino, who for almost five years, supplied the legally blind author the eyes, research, and computer skills that made this book a reality. Valerie Prôot and Mike Gagné, who as my French interpreters, translators and researchers, made my book their labor of love. Rebecca Roth, a retired English teacher, who knew how to pull my creative chain with a combination of flattery and criticism. Pete Noyes, author and investigative broadcast journalist, never hesitated when I asked him to put his work aside and evaluate my progress paragraph-by-paragraph, chapter-by-chapter. Marcia Meier's ruthless line editing transformed what had previously been a cumbersome, bloated manuscript. Because of his belief in the book, the late René Coté hand delivered old nightclub buddies for interviews and thus provided a look at the nether side of the CHAD story. Encouragement and priceless contacts were supplied by Martin Frassignes, General Manager of the former air base now Châteauroux Air Center. Lt. Col. Donna Hildebrand (USAF Ret.) provided a direct link to Air Force brass and high ranking Civil Service officials whose stories provided a rare insight into the Air Force class system. Dozens of French citizens overcame initial reticence and at times distrust to provide me with a collection of personal memoirs and photos. Their American counterparts had no such reluctance. René Valencia displayed uncommon patience teaching necessary skills to a sight impaired writer forced to retool his computer skills at the VA Blind Rehab Center in Tucson, Arizona. My thanks and apologies go out to others inadvertently omitted from this list.

CONTENTS

LIST OF PHOTOGRAPHS

PHOTOGRAPH SET C

FOREWORD

The history of Châteauroux is marked with the indelible imprint of the American presence. When the first American Army pilots came to La Martinerie Air Base in Châteauroux in 1917, a little bit before the end of World War I, no doubt they never thought they would be the beginning of a half-century love affair between Berry and the Star-Spangled Banner.

Fifty years later, when the Americans departed along with all other NATO troops from France, it caused real trauma in the city, both in spirit and everyday life. Since then, many things have changed. The pain passed with difficulty, the wounds still fester for some, but the century has changed. Châteauroux reconstructed itself and evolved, aspiring to finally become an important city in the new Europe.

Steve Bassett's book demonstrates that across the Atlantic, in a provincial town of France, the memory of this American period is very long-lived, proof of our undeniable love for this grand nation that is the United States of America.

However, there is always in the minds of the Castelroussins a nostalgia of those years in the heart of the twentieth century when they had gotten used to meeting GIs in the streets of our city, hearing American language in the villages of Brassioux and Touvent. And seeing for the first time these sports so strange to us that are the American football and baseball.

This nostalgia returns stronger every year in Berry, with recurring reunions that honor Uncle Sam. The friendship between our two countries is not ready to end when we see the enthusiasm and passion with which these reunions are organized. And it is always for us a pleasure to see reliving, even for only a few days a year, Châteauroux Air Station.

Jean-François MAYET
Sénateur de l'Indre—Maire de Châteauroux

INTRODUCTION

The bitching had been going on for days when the first trucks finally dropped into ankle-deep mud their cargo of American Air Force GIs along with their bundled tents and personal belongings. This sodden bunch cursed the missing sun and wondered if they would ever see it again in this bleak, backwater dumping ground called Châteauroux. They wondered what the Air Force brass had in mind for them in central France, but when it came right down to it, and as bleak as it was, it was a hell of a lot better than getting their butts shot off in an equally obscure place called Korea. Wake up, buddy, this was the Cold War.

If a nucleus of old-timers, but mostly kids from America's farms, small towns and urban jungles, were uncertain why they were in central France, this was not the case with the Communist cadres who surrounded them in early 1951 while awaiting their orders from the Kremlin. Throughout Berry, with Châteauroux in the center, Communist cells perfected the anti-American pamphlets, posters and street-corner rhetoric designed to blunt Uncle Sam's attempt to colonize this territory. A large number of the region's voting age natives were members of the deeply entrenched Communist party, and they were eager for a fight.

The United States would never allow Western Europe to go Communist. By 1950, Uncle Sam had the deep pockets and nuclear arsenal to make its plans a reality. That was the easy part. Implementation—now that was a different story. France was at the core of the solution and the problem for both Washington and the Kremlin, a juicy plum ready to be devoured by either side.

With the Truman Doctrine and the Marshall Plan already in place, America and its North Atlantic Treaty Organization (NATO) allies had already drawn the battle lines when disgruntled American airmen and Army engineers began arriving in Châteauroux in early 1951 to begin work on what was to become the largest Air Force supply depot in Europe. Nobody could blame them if they gave little thought to the fact that they were the vanguard of what was to become a global network of military bases positioned to thwart the Soviet Union's growing military ambition. During those early years in Châteauroux, the West's grand Cold War strategy was simple. It was reduced to keeping dry, warm and clean, hoping your next meal would be edible and that for another day you kept your boots from being sucked off into bottomless mud.

Fast forward to July 7, 2009, when U.S. President Barack Obama met with Russian Prime Minister Vladimir Putin and Russian President Dmitry Medvedev during a two-day meeting in Moscow that was hailed as "groundbreaking." This summit meeting underscored the United States' concern about Russia's punitive muscle-flexing around the world and elicited from Obama a cautionary admonition that, "The time has come to finally put the days of the Cold War behind us." He also warned, "We don't want to go back to a dark past."

Obama's conciliatory remarks provided insight as to why more than forty years of bleak and often bloody world history must not be repeated. His words nonetheless beg the question as to why anyone would write a book about a huge U.S. Air Force Base in central France that has been padlocked and largely forgotten since 1967.

Enough books about the Cold War have been written to sufficiently stock a good size library all on their own. Anni P. Baker's excellent *American Soldiers Overseas* and *Life in the U.S. Armed Forces* are studies of American forces at home and abroad with Germany, Panama, and the Pacific Rim outposts as the principle focus. France is given cursory attention, as though the sixteen-year U.S. military presence existed in a parallel universe. Any way you look at it, this was a long time. Considering the suspicions, jealousies, bigotry, and crass opportunism inherent whenever one foreign power occupies another, this book pieces together an improbable story that demands telling.

There is no better place to begin than at the sprawling truck stop just across the highway from the old base, where you can still see some of the most beautiful buildings in the Berry region of France. On any night, you'll find more than two hundred eighteen-wheelers jammed into the truck stop's vast parking lot. A potpourri of license plates from Spain, Germany, France, Portugal, Italy, and Belgium offer unwitting historical contrast to the white, art deco buildings across the road. The trucks, with their burly, heavy-handed drivers, represent the Europe of today. The base's architectural delights, with their three-story Grecian pillars, are silent historical reminders of a Cold War that now seems more fantasy than reality.

It should be noted that this book's title and subtitle are not meant to make light of the contribution made by American servicemen during the more than forty years that the Cold War dominated American foreign policy. For the more than 120 million men and women who served in the U.S. military during the Cold War years, it was a time of very real anxiety. I was among the 27 million military personnel stationed overseas.

From 1947 to 1989, the United States and its allies had become engaged in a new type of global war that might have been ideological in concept, but carried with it the threat of nuclear annihilation. Mothers, fathers, wives, sons, and daughters were sending their loved ones into a vast unknown. With the collapse of the Soviet Union, it all seems so unreal now, but for the kids we were sending overseas—the eighteen-year-old cowboy from Wyoming, the black sharecropper's son from Mississippi, the young steel worker from Ohio—their military duty would punctuate their lives forever. For those left behind at home, there were the abiding questions: Is our loved one safe? Is he or she in capable hands? And why the hell are they there in the first place?

During its sixteen years of operation, from 1951 to 1967, the Déols-Châteauroux Air Station would provide these answers in a fashion only rarely encountered at any other military installation in the world. Simply stated, it was a great place to be, safely situated as it was in a backwater region of central France. From Scandinavia to the Mediterranean Sea, all of North Africa and Turkey, every U.S. air base would depend on supplies shipped from there. The Châteauroux Air Station (CHAD) would economically lift Berry from its impoverished status to the point where ownership of the fabled, tiny, Deux Chevaux automobile replaced the bicycle as a status symbol.

If nothing else, this book defines irony. The base arose amidst the perceived threat of a worldwide Communist takeover. In 1951, McCarthyism had convinced us that Commies were everywhere. So why not construct this mammoth supply depot in the center of a French region dominated by the Communist party?

Golden Ghetto is not a totally objective historical treatise. It will not take long for history buffs to realize that this is just as much a collection of personal perceptions and experiences based on interviews that quite often led to friendships. If readers are looking for long lists of military hardware and how it was deployed, they should look elsewhere. Of course it will be necessary to identify the aircraft needed for CHAD to fulfill its missions, but I do so only when needed to advance the narrative.

This book is a collective memoir. I was lucky enough to find French, from the Left, the Right and the Center, who overcame skepticism of me and shared long-dormant memories. With the Americans, it was easier. They were a talkative lot, wearing their happy memories on their sleeves. When I embarked on this project, their numbers on both sides of the Atlantic were dwindling. Once they were gone, a remarkable story would disappear with them.

One aspect of the *Golden Ghetto* story that could not be ignored was the baggage both parties brought to the dance. Although this book is not chronological, it does attempt to provide insights into the historical, political, and philosophical differences of two disparate cultures. An examination of this force-fed, Franco-American ethos often revealed disagreeable and sometimes ugly realities that had to be dealt with.

My wife, Darlene, and I had the good fortune of being able to purchase a small country home just outside Bouges le Chateau in 2000. As our circle of friends and acquaintances widened, their stories, which at first sounded like fairytales as only the French could concoct them, eventually gave shape and substance to a massive air base my wife and I hadn't even known existed. Our home, La Cure, was one of eight structures remaining of the village of Ste. Colombe, about thirty kilometers from the old air base. During its heyday, the base gave new meaning to the phrase "living is easy." So easy, in fact, that the Cold War was only the pretext for a base where any thought of a shooting war seemed absurd. That is, until October 23, 1956.

On that day, thousands of Russian troops and tanks stormed into Budapest to quell a spreading anti-Communist uprising, and the world trembled. Throughout Europe, long-rehearsed emergency plans were activated at all U.S. military bases except for one. At CHAD they asked: "Evacuation plan? What evacuation plan?" Hardly anyone knew an emergency plan existed, although the base had been around for more than five years, and full dress rehearsals should have been conducted every six months. The plan was uncovered in an obscure filing cabinet, reluctantly put into effect, and once this passing annoyance was over, things returned to normal and would remain so for the next ten years.

Normalcy at CHAD, as it applied to Americans and Frenchmen alike, was a life where things only got better. Today, the French still bemoan the padlocking of the base after the dawn of the twenty-first century. And who could blame them? The French working at the base and those benefiting from their dealings with the American military suddenly found themselves enjoying a middle class existence they could never have imagined.

And it all centered around a military installation that easily met the criteria to support the apt title, *Golden Ghetto*. What in the world is a "golden ghetto"? It was the creation of an idealized American middle class existence on a barren farmland in rural France. People got to live in ranch-style homes with glittering appliances, big backyards, and neighbors you could trust and rely on. At first, my wife and I dismissed the stories of both the French and Americans as going beyond the fanciful to the far-fetched. For example, we were told that the vast, flat farmland surrounding our home was an ideal parachute drop area supplying Maquis Resistance fighters with guns, ammunitions, and explosives to sabotage Nazi occupation forces. We were told that Communist resistance fighters and their counterparts on the Right created their version of Tombstone's O.K. Corral, shooting it out in the open fields to see who would gather in the air-dropped supplies.

These stories gained support when we were shown photos after word got out that I was researching this book. We saw airmen, along with their wives or girlfriends, in full livery, mounted on sturdy steeds riding to the hounds. GIs, whose only encounter with sword play might have been an Errol Flynn movie, were taking lessons from a fencing master imported from Paris. Graffiti of "Yankee Go Home" and "U.S. Go Home" seemed to be everywhere, painted during early stages of base construction by Communists who would later hide their credentials when seeking work in the hangars and warehouses. A big paycheck trumps ideology every time. There was one French worker whose only job was changing burned-out light bulbs in the warehouses while earning more money than some professionals in town. The Marshall Plan put tractors and harvest combines in the hands of reluctant French farmers, some of whom preferred their horses instead and put the newfangled mechanical behemoths on blocks in their barns. If there wasn't a book somewhere in all of this, then my thirty-five years of journalism had taught me nothing. Yeah, there were the clichés, the black market stories, prostitution, French maidens ever eager to entrap naïve GIs in marriage, the ugly American with his big car, big bankroll, and even bigger sex drive, and Charles de Gaulle, who brought it all to an end. These are all part of the story.

When this book project began to take shape in 2004, Franco-American relations were in a dismal state. American tourism in Paris had dropped twenty-five percent in 2003. Americans were systematically destroying imported French products, especially wine, which was poured into the gutter in cities from coast to coast. It was misplaced reaction to French President Jacques Chirac's unyielding opposition to the war in Iraq. This made tantalizing headlines across the United States, as talk radio and television cable news commentators literally went crazy, while the Paris media was at its negative best.

Surprisingly, I found little evidence of anti-Americanism in Berry. In fact, in almost all cases, even among life-long Communists, there was emotion-laden regret that the base had closed. And why not? We had invited them through the front door of a Golden Ghetto whose glitter far transcended the boundaries of CHAD, offering an intoxicating allure virtually impossible to resist. For me, their stories—some sad and others exhilarating—created an irresistible magnet.

Golden Ghetto

"It's so easy to get trapped inside the Golden Ghetto. Don't be trapped inside the Golden Ghetto, get out and go into town, meet the people and discover the real world outside."

—Army Captain Francis C. Nollette's stern advice to his son, Frank Nollette, a student at the Châteauroux-Déols Air Station High School.

"The American presence was our spell of sunny weather in all respects."

—Leandre Boizeau, Communist publisher of the regional magazine, *La Bouinotte.*

"The Commies who painted the 'U.S. Go Home' graffiti were cowards. They did it at night, then ran home to hide."

—René Coté, former U.S. airman and later a Châteauroux businessman.

CHAPTER 1

TRANSFORMATION AND REACTION

Country girl, single, wants to spend life in the country. Seeks in view of
marriage farmer, estate manager, or similar. Send photo of tractor.
— Marriage Appeal in "Le Chasseur Français"

This appeal by a French maiden in the classified matrimonial ads section of "Le Chasseur Français" couldn't have summed up more succinctly what was happening throughout central France beginning in the late 1940s, a phenomenon that continued well into the 1960s. Under the protective umbrella of the Truman Doctrine, the French, as well as the British, Dutch, and to a lesser degree, the Germans, were reaping the benefits of Uncle Sam's deep pockets. It was payback time. Less than a decade earlier, America's vast war machine was a prime factor in eradicating Hitler's Nazism from Germany, a crusade that laid waste to much of Western Europe in the process. Enter the Marshall Plan, which along with the Truman Doctrine, was ultimately praised as the most successful foreign policy initiative in our nation's history.

Châteauroux, a town of some 40,000 people, owed its existence to agriculture. The rich soil of the Berry region was turned over at least three times a year for winter and summer wheat, the oil producing sunflower, and lentils. No longer the exclusive domain of plodding heavy-hoofed beasts, the energy was now encased in machines spewing acrid diesel fumes in their wake. The word "horsepower" now had a different meaning. John Deere, Caterpillar tractor, McCormack and International Harvester combines were now revered in the new agrarian pantheon.

In January 1947, Secretary of State George C. Marshall told an audience at Harvard University, "The farmer has always produced the foodstuffs to exchange with the city dweller for the necessities of life. This division of labor is the basis of modern civilization. At the present time, it is threatened with breakdown." It was the birth of the Marshall Plan.

Three months later, President Harry S. Truman told Congress, "The seeds of totalitarian regimes are nurtured by misery and want. They spread and grow in the evil soil of poverty and strife. They reach their full growth when the hope of a people for a better life has died. We must keep that hope alive." In the Berry region, with Châteauroux as its flashpoint, a strong and restive Communist cadre lay in wait. To Party members, Truman's words were a direct challenge. But a challenge to what? What did they want to protect?

Under the Marshall Plan, the United States would spend twenty billion dollars on strictly economic aid. There would be no military hardware included. This puzzled Josef Stalin. For him, there was always the quid pro quo that combined economic aid with military might. He publicly denounced the Marshall Plan as a "trick." Congress, never a body known for throwing around foreign aid readily, reacted swiftly to Stalin's denouncement and gave Marshall what he wanted. This massive infusion of money enabled Europeans to do many things, including broadening their marriage prospects. Here is another classified ad taken from Le Chasseur Français:

Land owner, 480 hectares, offers his daughter 22 for marriage.
Important dowry and farming rights.
Non-serious parties abstain.

While husband-hunting maidens were sending out man-wanted ads throughout central France without even a hint of embarrassment, it was a newspaper appeal of a different sort that prompted a cross-continent re-location to Berry by Elizabeth Reh and her three daughters. It gave Elizabeth that stroke of good fortune she had been looking for. The ad was circulated among the hundreds of Yugoslavian refugees of German heritage who were interned at Ernstbrunn Castle in southern Austria. Anna Reh was almost seven years old when her mother read the ad. Incredulously, her mother realized that she and her family were being

offered a prepaid ticket to freedom. In 1949, the Marshall Plan was fulfilling its promise in France, and French farmers in Berry found it hard
to keep up with the huge infusion of American dollars and equipment
that was dropped into their laps. The ad was very direct: "Farm help
wanted in the Châteauroux region of Central France. Transportation
and housing immediately available for Yugoslavian refugees." The Reh
family saga, particularly how a menial golden ghetto job would change
Anna's life forever, comes later.

In 2005, I interviewed my neighbor, Gerard Charbonnier, a wealthy
farmer with many hectares across highway D2 from our small home
in Sainte Colombe. For a young man then in his twenties, the Marshall
Plan ushered in profound changes: "It changed things. People changed.
We borrowed money and were able to buy tractors, American tractors.
At home we had a German tractor. We didn't have a Harvester. I remember our first one. You couldn't see what was behind you when you
were dragging it or when you turned around. We couldn't see the work
we did. There was a duck who had just laid some eggs in a hole. All we
found of her were her two legs."

Jean Diez, another neighbor who operated a driver's education
school in Châteauroux, was matter of fact in his appraisal of what he
saw as a young kid during the 1950s. "There were tractors. They were
American and the French used them. We had to follow the evolution of
agriculture. It was the survival of the fittest. Those who had the biggest
tractors and those who could spend less time in the fields were the ones
who survived."

But not everyone viewed this evolution through the same prism.
Charbonnier recalled a neighbor, now deceased, who stared in silent
wonderment when a John Deere tractor was off-loaded in his barnyard.
He told Charbonnier, "I've been using horses all my life. Sometimes
I had as many as twenty. Why do I need a tractor? I'll never use it.
I'm going to leave it for my son." Charbonnier said the tractor was
pushed, not driven, into his neighbor's barn and put on blocks, where
it remained until his son inherited the farm.

Recollections such as these were really memories of hope that embraced the belief that nearly a decade of despair had indeed ended. The

Nazis had come and gone, and the collaborationist Vichy government had disintegrated with the departure of its jack-booted patrons back across the Rhine. There seemed to be very little to worry about. And why should there be? A good-natured Uncle Sam had arrived with his bulging wallet. When developments were viewed from both sides of the Atlantic, the irony became apparent. On August 29, 1949, while smiling Berrichon farmers were putting their Caterpillar, John Deere, and International Harvester goodies to work in their fields, the Soviet Union successfully tested its first atomic device at Semipalatinsk, Kazakh SSR. Nuclear devices at the U.S.S.R.'s disposal led to elevated doomsday fears in America that would eventually lead the Office of Civil Defense to promote the digging of underground home shelters and issue handbooks on how to do it. Popular Mechanics, Life, and Time printed do-it-yourself instructions. The United States, relatively unscathed during World War II, was going underground. No evidence could be found that even a single Berrichon found the time or inclination to dig a bomb shelter.

COMMUNISTS EATING POPCORN

Lurking about in vocal opposition were the Communists. They knew that Berry, particularly Châteauroux, was in a sorry economic state. It needed help. So there arose that marvelous conundrum that provided the underpinnings for almost all Soviet theory—Hegelian dialectics be damned. Help was needed, but from the Americans? No way! There were murmurs that a major American Air Force base was on its way to Châteauroux. Whether the gossip was true or false, if the base were to become a reality, it would provide the most fitting testimony that the Truman Doctrine and Marshall Plan had arrived. There were tractors, harvesters and American trucks on the ground and U.S. war planes in the air providing protective cover so the wheat and lentils reached market. As one observer noted, "How could the Commies beat that?" Soviet-inspired, anti-American rhetoric and demonstrations would be reduced to harmless street graffiti. The Communists of that era are older now, but their memories linger.

When I went to the comfortable, well-appointed Châteauroux home of Pierre Pirot, a Communist for more than sixty years, I found a jovial, smiling man of seventy-seven, ever observant and preoccupied with whether or not your wine glass was full. He recalled, "Most of them [Communists] who had been resistant considered that the country had just recovered from an invasion, a Nazi occupation. The people of France didn't want a new occupation even if it came from the United States."

I was at the home of my interpreter and good friend, Valerie Prôt, in 2005, preparing for my interview with Leandre Boizeau, publisher of the glossy, regional magazine, *La Bouinotte*, when I discovered one of the new faces of French Communism. Boizeau arrived in a shiny, silver Mercedes-Benz convertible that appeared to be half a block long. Valerie's husband, Benoit, didn't think he had ever seen a car like that on his street before. He estimated that it cost about 75,000 euros. It was more than fifty years ago when a young Leandre took to the sidewalk with propaganda supplied by his parents. "I was eight or nine years old. I belonged to a Communist family and I remember walking in the streets with petitions against the U.S. intrusion in France," Boizeau remembered. "It was the Communist doctrine, since they felt they would be kept from power. They understood they would be isolated on a national point of view as well as on an international basis. Everything that was said was against the U.S.

"I was still a kid, eleven years old. When the Americans would pass by in the streets they would toss us chewing gum. We would greet them so they would give us gum. For us, the Americans were like big kids, nice.

"We got to know them very well. Many of them lived off the base. There were lots of Americans who rented houses. And so there were neighborly relations, which were very nice because we invited each other over. Some Americans came over to our house for dinner, and we went over for dinner as well. It was pretty strange. They even invited us to watch movies on the base with their bags of popcorn, which they ate. That creates special bonds."

For Boizeau and his family as Communist acolytes, it was a losing battle from the beginning. The petitions might well have contained the very best in hard-hitting, anti-American rhetoric, but how could they beat Charbonnier's matter of fact logic? "We never went hungry. That was the point of the Marshall Plan," he said. "It was intended to help us."

These were the strongly expressed feelings, pro and con, regarding Americans as France moved into the 1950s. Memories of the American GI as the liberator were still fresh, but by 1950, the French were asking, "But do we really want them around?"

A New Base and a New Breed

It was raining when U.S. Air Force Brigadier General Joseph H. Hicks got into his unmarked military staff car in Paris in February 1951. It always seemed to be raining that winter. This day was no different, as the rain continued for the duration of the three hour drive to Châteauroux. With him were three aides. Earlier that year, the French government had agreed to turn over La Martinerie, an old French air base dating back to 1917, to the Americans. Plans called for the base to eventually be the chief supply depot servicing the air forces of the North Atlantic Treaty Organization (NATO) countries. It would be Hicks' first visit to La Martinerie and he frankly admitted he didn't know what to expect.

At fifty-two, the Glenwood, Mississippi native knew his military career was aging gracefully toward its end. The West Point graduate got his wings in 1925 and, looking back over the past twenty-six years, he accepted that he had never been a hot-shot flier or bomber pilot. Moving from air photography through assignments in supply and procurement, he had carved a niche for himself in the Air Force, which became increasingly valuable as Cold War demands changed the face of the U.S. military. Châteauroux was a plum assignment. Once in place and operational, it would provide an important supply anchor throughout Europe, North Africa and the Middle East.

As his car sped through the outskirts of Orleans, continuing south along N20 toward Vatan, Hicks wondered about the young enlisted men who would be serving under him. The Korean War had erupted the previous June, catching the Truman administration by surprise and totally unprepared. The Chinese Communists had amassed a huge army and, in an unfathomable failure of U.S. military intelligence, had penetrated undetected deep into North Korea, encircling and essentially liquidating the U.S. Army's Second Infantry Division. The Army and Marines needed young bodies, lots of them, and right now. They were to be cannon fodder, pure and simple. The panic that careened up and down the halls of the Pentagon was palpable. At first, there were going to be six weeks of training in the States, but that was downsized to four weeks and then two weeks.

Finally, underequipped, poorly trained, and scared, these kids were dumped into the shrinking Pusan perimeter. Savvy reporters soon had this story out on the street. Local draft boards were pumping out notices as fast as they could and the trade-off began. Faced with what could be a deadly two years in the Army or Marines, more and more eighteen-, nineteen-, and twenty-year-olds found a four-year enlistment in the Air Force suddenly inviting. Many of them were coming to Châteauroux. But with what kind of commitment? Hicks had carried with him an abiding love for flying and the military service which had made it possible. It would be interesting to find what kind of young men he was getting.

One of these young airmen, among the first arrivals in 1951, was Gene Dellinger. A Newton, North Carolina native, Dellinger was a healthy eighteen-year-old kid fresh out of high school.

"I joined the Air Force because my brother had been drafted into the Army," Dellinger said in June 2009. "I was ready to go to college and had been accepted, but I knew my draft number was due to come up pretty soon. My brother told me, 'join anything, join up but don't allow yourself to be drafted.' And sure enough, my draft number came up shortly after I joined the Air Force."

Hicks, as the first CHAD commander, and Dellinger, as a wide-eyed kid away from home for the first time, found themselves in a region of Berry that had been relatively untouched by World War I, but had been the target of frequent bombing raids during World War II. Photos taken in the aftermath of the Allied bombings showed complete devastation. Hangars, machine shops, maintenance buildings and warehouses were leveled. A German Focke-Wulf 190 fighter plane could be seen partially buried in the debris.

As part of the Free Zone governed by the Vichy, the region suffered wartime hardships, including destruction of the Châteauroux railroad station and electrical power plant, as the result not of German, but Allied bombers. For a while, the Germans had taken over the airstrip at La Martinerie and during the closing months of the war, German ground forces moved through the area as the Vichy government collapsed. Now there would be a massive buildup of American forces, a military occupation never before experienced in modern Berry. How would the citizens handle it, having never had to cope with anything like it before?

Hicks' thoughts drifted from what would be awaiting him in less than an hour as his car splashed through Vierzon. His thoughts wandered back to New Guinea and the horror he had found at the three Hollandia airfields, strangely more real now than in 1944. He had been named supply chief for the Far East Air Command, arriving shortly after two years of bitter fighting that tested human endurance. The stinking jungle, swamps, ungodly heat and humidity, up to 300 inches of rain a year, malaria, and jungle rot all provided a horrible landscape for human slaughter. Scattered about the surrounding jungle were the skeletal remains of the Japanese enemy. At the airstrips, the carcasses of many of the 340 enemy planes destroyed on the ground remained untouched, mute, rusting testimony to the Japs' hopeless dream of empire. Hicks had frequently thought of this cesspool as symbolic of a backwater campaign that was little thought of and even less appreciated back home in the States.

THE UGLIEST CITY IN FRANCE

When Hicks' car pulled to a stop in front of the St. Catherine Hotel in downtown Châteauroux, emerging was a complex man about to undertake the toughest assignment of his career, an officer who would never forget the past, but realized that these new responsibilities would require him to live in the here and now.

Hicks was not a big man, physically. He was short, with a gruff demeanor that would serve him well as he tried first to assess, then to lead a new generation of airmen through what would prove to be three years of chaos. "Hicks was a knock them down, drag them out kind of a leader," Dellinger said. "He commanded respect. Hicks had the biggest star I had ever seen on a general's car in my life. Maybe it was because he was small. I remember one time there was a group of us standing at the gate at the air base when he drove in his Chevrolet staff car with the big star on it. We were all looking the other way so we didn't salute him as he drove by. Hicks jumped up on the back seat, and I'll never forget it, he stared at us while ordering his car to stop. Hicks asked 'who's in charge here?' And because the two staff sergeants were ranking NCOs, they were the ones who got busted and both of them lost their stripes because they hadn't saluted."

And for the first month, the "here and now" would be realized in secret. With his aides and an interpreter in tow, he would visit La Martinerie and an underused airplane factory, planning to prepare two airfields that would become the Châteauroux-Déols Air Station. Because no explanation was given, the townspeople were kept in ignorance until March 19, 1951, when the local newspaper, *La Nouvelle République du Centre Quest*, reported that up to four thousand U.S. military were on their way. Who would believe such a preposterous thing? Here in Châteauroux? In the middle of nowhere? An American air base? The blinders were off and the debate began. It wouldn't end for sixteen years.

And it wasn't as if the local citizenry had joined forces to safeguard one of France's tourist Meccas. In fact, Berry, and the way its inhabitants handled the French language and behaved themselves on a day-to-day basis, were derisively referred to throughout France as "Berrichon." I remember several occasions when one of my Parisian friends would encounter a provincial hayseed who had a phonic lapse or two or perhaps improper tongue placement. "Berrichon," would be my friend's reaction and it didn't make a bit of difference if the linguistic transgressor was from Berry or Brittany. Châteauroux was then, and remains, the heart of Berry.

"Oh Châteauroux, the ugliest city in France," said Jean Giroudoux. "I accept all your side streets, I turn you upside down, I disarrange your hair, I love you."

Giroudoux's flamboyant description was based on experience. He was a student in Châteauroux from 1893 to 1900 and later gained international fame as a playwright for his *The Mad Woman of Chaillot* and *The War of Troy Will Not Take Place*. The Berrichon know full well how it is possible to love even the ugliest and most critical among them. After all, when a man of Giroudoux's renown calls you the ugliest town in France, it has to be worth something. And the good citizens of Châteauroux rewarded Giroudoux with a *lycée* that has borne his name since 1949. Today, if you look closely, you will find a plaque somewhere in the Old Town area commemorating him.

Two years after the school was named for Giroudoux, a small caravan of army trucks wound their way from the downtown railroad station to La Martinerie. For Airman Nick Loverich it was the end of an odyssey that began in the States. The train ride from Paris to Châteauroux proved to be foreboding of things to come. "We arrived by train after traveling third class, which they still had at that time. It was crowded and not very comfortable. When we got there you could see that it was an old base, originally from World War I. First we waited in line for three hours in order to eat at this mess hall set up in one of the old hangers. There was mud everywhere. Everything came in trucks. One truck dumped off our personal belongings, and another dumped our bundled up tents. It didn't matter where because everything landed in the mud."

Loverich, at the age of twenty-four, had little in common with most of the nineteen- and twenty-year-olds who arrived with him. Unlike them, he had been shaving for at least seven years. Loverich was born and raised in the steel town of Youngstown, Ohio, one of thirteen kids. He quickly learned to appreciate what his steel worker father had endured in raising his family. Loverich joined his father at the mill while still a teenager and it didn't take long for him to decide that a life of tending huge crucibles of molten steel was a soul destroying existence he had no intention of enduring.

"I was eighteen when I joined the Navy. I stayed in for two years, 1945 and 1946. In 1951, after the Korean War broke out, I was recalled and that's when I enlisted in the Air Force," said Loverich, who was eighty when I interviewed him in the Autumn of 2007 at his lovely home in Issoudun, roughly halfway between Châteauroux and the historical city of Bourges. Loverich, at an early age, discovered his gift for drawing. The Air Force quickly put this talent to use. He found it interesting that as a member of the Graphics Presentation Branch he was among the first airmen to arrive at La Martinerie. The Korean War had been raging for a year and the Communist threat, once theoretical, was now real. Although the Soviet-supported partisan threats in Turkey and Greece had been contained, there were other potential hot spots in Europe. It was no secret that Stalin considered France the biggest prize in Western Europe. In 1951, the propaganda war was in full swing and Loverich's skills made him an excellent propagandist. "Our branch was responsible for all graphics and pictorial projects, including posters, pamphlets and booklets to be distributed to our forces and among the French. These included anti-Communist slide shows presented at schools." With the stakes so high, it was never too early to get your message out.

Marseillaise cranking out anti-American diatribes at a furious pace. It took less than six months for its writers to get up to speed with headlines such as these:

10/17/51: *Châteauroux taxpayers must provide $500 million to accommodate the Americans as (French) workers are crammed into barracks.*

10/20/51: *Posters in English that are subtitled in French begin to cover the walls throughout Châteauroux.*

10/23/51: *Châteauroux is an occupied city. Nightclubs, dance halls and hotels are shady business zones of gold.*

It is doubtful that the propaganda war made much of an impact on Air Force GIs and Army engineers who endured without basic creature comforts. They were a forlorn bunch of guys against whom all the forces of nature seemed to be conspiring.

The job on the surface seemed simple enough—if you could get to it and that was no easy thing. "The mud was really troublesome. I'd say it was between twelve and twenty-four inches deep. It sucked the boots right off your feet. We had to build wooden platforms to walk on, but we couldn't cover all the mud. There were even deeper hidden sinkholes all over the place," Loverich recalled with a distasteful shrug.

"I had a friend whose name was Short. Really, I'm not kidding. And the name was appropriate because he was very short, only a little over five foot. Going for chow one night, we had to walk through the mud and on the way we lost Short. We looked around for him and discovered that he had stepped into one of the deeper sinkholes. He was trapped, buried deep in the mud. We had to go back and pull him out. It took three of us to do it, leaving his boots behind," Loverich smiled with the memory. "We cleaned him up, but he never did eat that night."

Tent City, Quon City, or Just Plain Mud City

It was hard to imagine that the quagmires at La Martinerie and at the air base being constructed two miles away at Déols could get worse. But constant rain thwarted any attempts at optimism and raised pessimism to new heights. It appeared that Mud City was there to stay. It took thousands upon thousands of sodden man-hours of back-breaking work to show progress. Rain, mud, sinkholes be damned; there would be a "Tent City." Spread across the southeast end of the base, the original 240 tents with ten beds each were eventually installed. It was primitive housing in which the only heat was supplied by a potbellied stove.

Tent City eventually disappeared, giving way to row after row of Quonset huts aptly named "Quon City." Permanent buildings followed. Many of them are still there today, most of them empty and an eerie reminder of an epic transformation more than half a century ago. But in a very real sense, Mud City still remained more than fifty years later, and if you were to stray off paved or gravel roads or pathways you would quickly see that it would be a good idea to make sure your shoes or boots hadn't been sucked off.

Mud was an unwanted bequest that never diminished. Until permanent runways were installed at Déols, equipment sank into the mud, the heavier the deeper. When a C47 transport landed with supplies, it couldn't be left in place overnight. The next morning, cranes would be needed to lift it out of the mud. But in light of the Communist threat, the equipment and material kept coming. In July 1951, Captain Jack Warren and six others composed the first contingent of pilots to guide their C-47s onto the wet, marsh-like, grass landing strip of La Martinerie. Warren, a holdover from the old "brown shoe" Army Air Corps, found conditions primitive at best. "Now like I said before, at the airstrip which was all grass, I could have taken off at any time with my wheels still locked because it was always wet and slimy. Instead of rolling down the field, you slid down the field. Mostly we were bringing in personnel, not heavy freight. At that time, because things were so bad, most of the freight came in by rail."

Besides flying an average of three missions a week, Warren also served as the administrative officer for the food service squadron, a job that would bring him to the attention of Hicks. Warren's duties included supervision of the make-shift mess hall in an old French hangar for enlisted men and officers alike. It did not take him long to discover just how insatiable the mud could be.

A large and heavy ice cream making machine, delivered to the mess hall one evening and left outside overnight, had sunk into a grave of brown goo when the cooks arrived the next morning. The ice cream maker was a gesture by Warren to raise the level of mess hall conditions from miserable to barely tolerable. "It was always cold there, and the meals were served on metal trays. In order that your food did not immediately get cold once your food was on the tray, they had set up trash cans, filled them with water, and had a fire going under them all the time and this heated up the trays," Warren recalled. "I can tell you that there were guys who ate their meals with their gloves on, that's how bad the conditions were in the mess hall.

"We had plastic cups for coffee and the silverware was just thrown out in trays. So you can imagine the theft that went on. There were guys taking the silverware to their homes, which they were just moving into, so I went to my commander and told him about it. I said the theft is so bad, I'm going to need more supply of silverware in order to feed the men.

"My complaint went all the way up to a command meeting. General Hicks was there, and I was told by one of the command officers, that Hicks said, 'Who is the guy in charge out there? Can't he handle it, if not get rid of him.' [Laughter]

"I kept my job, a job I really didn't want because after all I was a pilot, but I still worked very hard at it. I also had to accept and turn my back on some unpleasant facts. I was certain there were a lot of officers taking things, too. Look at it this way, you're setting up living quarters, so what's wrong with taking a knife, a spoon, a fork, a cup, and whatever you need. Yes, they were stealing, but it wasn't to make a profit, it was to set up their new home.

"There was also the black market to deal with, especially as it involved our French kitchen and mess hall staff. There was a lot of stuff taken from the mess hall, but I don't know if I could say it all went on the black market. I'm talking about the cooks and women servers who would help in the kitchen. They would take turkeys, and chickens, and so forth out of the big freezers that we had. They placed them under their smocks, and then carried it out when they left at night. I don't think they did it for the black market, I think they did it to survive.

"This was a poor region enduring tough times," Warren said. "We discovered that they would dip the frozen chickens or turkeys into vats of hot butter that was used by cooks for much of their cooking. The butter would harden almost immediately, long enough for them to get home before it melted."

Eventually, Warren's precarious take-offs and landings were made easier by the installation of Pearce planking. The planking was composed of corrugated metal landing mats that could be pieced together to provide a rudimentary landing surface and it helped to keep first the C-47s and later the heavier C-119 cargo planes from sinking into La Martinerie's muck and mire.

It seemed that everything was happening at once, tents going up, heavy equipment sinking and then being retrieved from the mud before it could be put into operation and all the while there was an emerging tent city. "During the Tent City days it was really miserable. If you had a good pair of boots to cope with the mud, you were lucky. Even the showers, you were lucky if you had one of them a week," former airman René Coté said. He was among the second large contingent of GIs to arrive. "Cars and other vehicles were stuck in the mud and were hard to get out. Even the base commander stayed away." When General Hicks did show up, his visit was generally unannounced, carrying with it a sense of foreboding. Dellinger recalled one such visit. "There was a time at our tent at La Martinerie. We always tried to make it look as best as we could, even planting flowers. We were outside the tent looking down as we stamped the mud off our boots. At the same time, Hicks made one of his surprise visits. We were standing in mud puddles when he stopped his car and got out and shouted, 'Who's in command?' It happened to be a black tech sergeant, a really tall, friendly guy who Hicks called over. We were all ordered to get into formation and march behind his car down the muddy street with Hicks and the sergeant marching in front of us. We were all marching in step through the deep mud, it was pitiful.

"It was easy to see why Hicks was considered a martinet. He could be a danger. But at the same time, he was a good general because he was keeping people together at a new base. He felt that even under the worst conditions he never wanted us to forget we were in the military."

Nick Loverich always believed that commanders shared both the good and the bad times with their men, not every day mind you, but certainly enough to let the guys know that you understood their plight and if they kept their chins up, things would get better. However, he didn't think that this timeless axiom held true with General Hicks. "First, he wasn't the friendliest, he didn't talk too much. It doesn't mean he had to chat with us every day, but we hardly saw him," Loverich said. "Here is an example of what he was like. We were still living in Mud City when he called for a full dress inspection. This meant we had to stand inspection in the mud, in our dress blue uniforms. This didn't go over very well. Inspections take a fairly long time, and the longer we stood in the mud, the deeper we sank."

Another example of how Hicks ran things was the state of the showers, which were actually just showerheads welded onto the bottom of fifty-five gallon fuel drums that had been filled with water. "It was a luxury to have hot water; we hardly ever had any," Loverich said. "Then, one day, the base was inspected by a congresswoman from Washington, D.C. I can't remember her name. Suddenly there was plenty of hot water for everybody. But the moment she left, there was only cold water. We appreciated the hot water for the few minutes that we had it, but we didn't appreciate that it was only for a few minutes while the congresswoman was at the base."

This was a harsh introduction to France for young Air Force recruits, because they didn't want to live in tents or sleep on the ground and were willing to devote two extra years to avoid doing that. Sharing a potbellied stove with nine other men in his tent and cloistered by the elements, the average airman began to wonder if there was another France out there somewhere, and when the rare opportunity arose, sloshed into town to see for himself. He quickly learned that there was a *cordon sanitaire* protecting Châteauroux's bars and restaurants that could only be breeched after he swiped the mud from his boots.

With truck after truck off-loading foreigners day after day, at first forty or fifty at a time, and then by the hundreds, a restless curiosity spread throughout Berry. Another invasion? How could it not be? Much to my surprise, many of the people I interviewed had never even seen an American, except in the movies. There were those who had remembered him as a liberator, but that was seven or more years before. The average Berrichon took pride in happy isolation, being off the beaten track and unmindful of what the outside world thought.

Teenage Berrichon were ambivalent about the ever increasing American presence. Girls were generally wary but welcoming. For good reason, teenage boys viewed well scrubbed GIs as threats to what would have been an easy transition to manhood.

GOOD OLD AMERICAN KNOW-HOW

"My girlfriends and I at the *lycée* had heard that the Americans had begun coming into town and we were dying to see them. But being proper girls, attending a proper school and with strict parents, we were forbidden to have any contact with them," recalled Lillianne Diez, who was fifteen when curiosity finally got the better of her. She eventually married an airman and was sixty-nine when I interviewed her in her El Paso, Texas home. "To approach them on the street would be unthinkable and we knew that if our parents found out we would be punished.

"After that, to say we met an American would be overstating it, let's just say we looked at Americans," Lillianne said. "We were told that they brought their laundry into town and there was one cleaners near the *Marie* (city hall) where we could see what they looked like. We would never think of going alone, so a small group of us went on this forbidden trip.

"The first thing we saw were guys who were taller than most French men, cleaner and good looking," Lillianne said. "Remember, we were a bunch of scared girls who wanted to bring back some real juicy gossip for our classmates the next day."

Without her realizing it, Lillianne's stealthy pursuit of juicy gossip provided fleeting acknowledgement that Châteauroux would never be the same again. Americans were different, as was the world they were creating in Berry. It was to become a glittering encampment, first defined by Frank Nollette, a CHAD High School student during those early years. Frank identified the ingredients necessary to create a Golden Ghetto.

"I have my father to thank for the advice he gave me. 'It's so easy to get trapped inside the Golden Ghetto. Don't be trapped inside the Golden Ghetto, get out and go into town, meet the people and discover the real world outside.'

"That was the first time I had heard the term, 'Golden Ghetto.' It meant that basically everywhere Americans went they set up a little ghetto that was made as American as possible. Everywhere I went throughout Europe, Turkey, and any place that Americans went overseas, the military—especially the dependents—only saw and interacted with what they found inside the ghetto. The only contact they had with the locals was with those who worked at the PX, the commissary, the clubs and other facilities at the base. Or with the little cottage industries which surrounded the base and were in direct support of it.

"Châteauroux definitely fit this description. Any place the Americans set up a military base it was the same. Even if you went to a remote place, you will always find the PX, the commissary, the officers' club, enlisted men's club, the bowling alley, the gymnasium. We create little pockets of America throughout the world, places where Americans could be very comfortable without ever going outside.

"Strangely, I did not detect any real jealousy among the French in Châteauroux because of the wealth within the ghetto. The French, being French, had no problem adopting a phrase used by the British during WWII when they described the American GI in England, 'over-paid, over-sexed, and over here.' Once you got into town you got to know them. If you were an American family—the military guy, his wife and two kids living on the economy, trying to make ends meet, buying your groceries locally—you got a very different impression of the French, and they got a much different impression of you.

"Living on the economy followed exactly what my father was talking about. You lived off the base in town and had to pay local utility rates, rent and had to deal with the exchange rate every day. You went to market, bought your bread and other staples from the French. You had moved out of the Golden Ghetto. The French came to realize that the Americans weren't all pariahs, that we weren't all 'ugly Americans.' And believe me, there were an awful lot of ugly Americans to spare."

The CHAD Golden Ghetto was no more than a drawing board concept that seemed almost unimaginable during those first few years. At La Martinerie, the drudgery inevitably provided outlets for good old American "know-how" and ingenuity. For René Coté—easily described in American parlance as an "operator," but in France admired as an *entrepreneur*—being savvy during hard times came easily. Coté never fully explained whether the following enterprise was legal or not. First you had to find an unwitting accomplice, in this case a tiny, family-run bar not far from the base. "We were able to go off base to buy cheap French wine by the crate," Coté said. "We'd buy two crates with eight, one-liter bottles in each. The wine was cheap but it was good."

Coté transformed a potbellied stove into a still and used the wine to produce an eye-watering elixir, a cure-all that he insisted fifty years later should have been patented. "We'd put the bottles around a pot-belly still to warm the wine. We added sugar, cloves, cinnamon and lemon peels, when we could get them, to the bottle to spice the wine up," he said. "We found that our toddies had very good medicinal value. The guys who made their own hooch were sick a lot less than the guys who didn't. We called it GI Gin."

The base had hardly opened when it began to prove its worth as a marriage broker. Witness the experience of Nick Loverich. The damp and dirty drudgery of that first year led him to pull out an appliance that made him not only a lot of friends, but some very serious money as well. Prior military experience proved invaluable.

"I had been in the Navy for two years and had learned to have an iron with me at all times so my uniforms would be pressed. When we finally got electricity at Tent City, I pulled out my iron to press my uniform. Another guy in the tent noticed and soon the word got around," Loverich said. "I began charging one dollar for this and a dollar for that. The GIs knew they had to come to my tent if they wanted a pressed uniform."

With his booming business in place, there followed events which for Loverich, almost sixty years later, still seemed like a fairytale. "It was kind of peculiar. After the first three months at the base, living in tents and mud, I went with my master sergeant when he invited me to accompany him on a drive to see the countryside. Just something to get away for the day. We came to Issoudun. While driving down a main street, I looked up and saw a girl standing on the balcony. She looked very, very beautiful to me. You could say it was like a fairytale, Romeo and Juliet. It was something that was meant to be. It was my first time in Issoudun, and I learned later that it was her first time back home from Paris where she was going to school. I smiled and waved to her, and she smiled back and waved. It was fate, meant to be."

But fate unrequited would have been a terrible misfortune for Nick, and the only possible way for him to see Françoise on a continual basis would be by car. There was no public transportation. Coming to his rescue was his beloved iron.

"I needed money to buy a car. I found that there was a 1946 Ford for sale. So for three months I basically stayed at the base pressing clothes in my free time and I ended up making enough money to buy the car."

And what followed was a marriage of almost sixty years with Françoise, a daughter, Corrine, and a lifetime in France that has given Loverich regional celebrity status.

Back to the mission of the base. Enter the Communists. First, I must say that it was hard for me to find more than a few bona fide Communists while doing my interviews. Hardly unreasonable—after all, it was 2005 not 1951, when one out of four of Berry's electorate proudly displayed a red rose in their lapel or on their blouse. Dining and drinking establishments were festooned with red-topped tables and crimson-bannered entrances. This was reason enough for a bunch of young, clean-scrubbed kids recently shipped over from the States to temporize that maybe Joe McCarthy was right—there were Commies everywhere.

"I had heard that the new recruits, before leaving their classes in the States, were briefed that all French men were Communists and lazy drunkards and women were all prostitutes," said Maurice Jaunot. "Little by little, the GIs found out this was only half true. We don't all drink," he joked. Jaunot was one of the skilled aeronautical technicians working for the National Southwest Aeronautical Construction Company (SNCASO) when the Americans arrived.

There were differences in language and work habits as both Americans and French struggled under the shadow of the big red bear that roamed the factory floor every day. Jaunot remembers, "A friend of mine, Jean Luneau who used to hang out with the GIs, had an American co-worker who would ask him every day, 'Hey Jean, do you know what a Communist looks like?' After a while, tired of his daily questions, he took the airman to the back of the shop and told him, 'See that guy, he's a Communist, see that other guy, he's a Communist too and by the way, I'm also a Communist.' After that, he never asked Leneau again and we laughed about it."

But for the hard-core Communists it was not a laughing matter. Within weeks of the first arrival of Americans, the "U.S. Go Home" and "Yankee Go Home" signs and graffiti began appearing throughout Châteauroux and Déols, as well as in the surrounding countryside. For the tens of thousands of impoverished and out of work Berrichon, the sentiments conveyed by these signs made little difference; they were hopeful that better times were ahead. And maybe the Americans would provide them.

1951 – COLD WAR POTPOURRI – 1952

- In May 1951, U. S. Army Intelligence ended weeks of foreboding speculation by confirming for the first time a massive infiltration of Chinese communist troops into North Korea.

- The black and white barrier in network TV was broken in June when a CBS variety extravaganza was televised in color.

- In June, General Jean de Tassigny's French forces routed the communist Viet Minh and eliminated the enemy pockets in the Red River Delta in North Vietnam.

- Mickey Spillane cemented his reputation as the king of racy crime fiction with the publication of two more sex-laden novels, *One Lonely Night* and *The Big Kill.*

- *Mad* magazine makes its debut in May 1952 as a 32-page comic book filled with nonsense with a message, and madcap satire, preparing the way for iconic Alfred E. Neuman in issues to come.

- On July 11th, Cleveland disc jockey Alan Freed invented the phrase "Rock-n-Roll" to put a less objectionable face on rhythm and blues, or so called "black music," featured on his radio show *Moondog House.*

- 1952 Presidential candidates, Republican Dwight Eisenhower and Democrat Adlai Stevenson, unveiled the nation's first political TV ads. Stevenson's ad encroached on *I Love Lucy*, angering millions of fans.

- On November 1st, a month after Britain's successful test of its first atomic bomb, President Harry Truman announced that the first hydrogen bomb had been successfully tested at Eniwetok in the Pacific.

CHAPTER 2

PENCIL PUSHING PILOTS

Every Sunday evening, the department heads at the base waited anxiously for phone calls to find out if all of their French employees would be returning to work on Monday morning. You could blame it on the Deux Chevaux.

—JAG Attorney Robert Goggin

Bob Goggin is referring to an all-too-frequent consequence of a uniquely French auto insurance program that the military sardonically dubbed the "Cinderella Insurance Policy." Perhaps it would be a good idea to describe what the *Deux Chevaux* (2CV) was. The literal translation from the French *Cheval Vapeur* is horse steam, which in turn means horse power.

Goggin served for almost three years as a lawyer for the Judge Advocate General (JAG) at the Châteauroux Air Base. Accidents involving big, high-powered cars driven by Americans and 2CVs driven by the French could mean a lot of paperwork for the JAG. One can easily imagine the mismatch when a 1957 Chevy convertible with its powerful V8 engine even nudged a *Deux Chevaux* on a country road in Berry.

"You know the French were terrible drivers. It seems they were getting killed on the roads every day. On top of that, they drove these tiny little cars, the *Deux Chevaux*. They were like little tin boxes on wheels with motors in them. The motor had only two cylinders, two horse power. It seemed that every Frenchman had one of them because they were so cheap and didn't burn much gas."

The 2CV might have been "a tin can on wheels" but it was also a strong indicator of what the base was doing for the region's economy. The mission of this mammoth air depot was simple: to supply all that was needed for every Air Force base on three continents. For hundreds of French workers, one dream of a better life included a first car.

With ample cash in their pockets, more and more French could purchase these little four wheel trophies.

"These tiny cars were very cheap, but the insurance premiums were very high," Goggin explained. "Many of the owners couldn't pay the premiums. So the authorities that handle matters like this came up with an idea. They wanted the owners of the *Deux Chevaux* to enjoy their cars so they instituted a unique insurance plan. They said they would give the owners an insurance policy that began at four p.m. on Friday and ended at midnight on Sunday. That's why we called it the 'Cinderella Insurance Policy.'

"The car owners were able to enjoy themselves for the weekend, but then we found that they were killing themselves well above the average rates because they wanted to get home before midnight when the insurance policy ran out. If they got home after midnight they weren't covered.

"I'm not certain exactly how long the program lasted, but it was about two or three years. They had to discontinue it because of the reckless driving. I know this sounds surreal, but can anybody make that up?"

This assessment of roadway mayhem by Goggin, who arrived at CHAD in 1962, was made roughly a decade after local Communists let loose a salvo of anti-American propaganda, both in print and with wall side graffiti. The attacks centered on two Communist perceptions: the image of arrogant, ugly Americans, and the big, powerful cars they drove in order to preserve this image. Two articles in *La Marseillaise*, the regional Communist newspaper, read like traffic accident reports with a bad attitude.

The first one appeared in May 1952, and is included in its entirety as a barometer of how far anti-Americanism could go if given an outlet. "A U.S. driver, who was completely drunk and driving above sixty-two miles per hour, mowed down and killed a father who had three children near Niherne. The population demands exemplary punishment for this Yankee after this fifth fatal accident.

"Last Friday evening, Mr. Uridat was coming back home on his bicycle. He was on his right hand side and was suddenly thrown away by a

Citroën TA, driven by a Yankee, a reckless driver who went on without caring until the place where he lives in Chateau de Surins near Niherne.

"The witnesses pursued the road hog who was driving so fast that they didn't manage to reach him. The Villedieu police finally stopped the man who remained in his car.

"The victim's friends showed their hostility. The man who was in the car put his hand into his pocket which is a gesture that perfectly proves the reason why they are in our country, ready to make new corpses and so to satisfy their tendency to kill people by thousands: a tendency that we do know very well.

"John Hazelip (it is his name) is of medium height, he's blond-haired, his complexion is sallow, his eyes are half-closed, his clothes are untidy and he has a nonchalant attitude. He has the typical attitude of a perfect reveler.

"Indeed, since Hazelip has lived in le Chateau de Surins, it has become the place of various types of binges. Driving drunk most of the time, Hazelip is dreaded on local roads. He has already had a serious incident on this same road.

"After this fifth fatal accident caused by Yankees, and even if US authorities try to minimize the problem, this man must stay in jail and be seriously condemned. It is necessary to underline the fact that if Yankees look cautious, it is fake. Even if downtown they still respect some road signs, when they are unleashed they drive with no restraint."

Another article in *La Marseillaise* appeared in December 1952. "On Place Lafayette, a front-wheel drive Citroën was going towards rue Victor Hugo. Suddenly, a U.S. car [with license number] SSTT2X bumped into the rear of the car. Nobody was injured, but the rear bumper and the trunk were damaged. A sworn official came. The Yankee despised the questions coming from the man representing law. It was necessary for the persons who were present to be firm so that the Yankee could be calmed down and recover some sort of quietness. This is more new evidence describing the civilization of these Yankee invaders."

Platoons of front-end loaders burned rubber as they constantly moved from warehouse to warehouse, to incoming and outgoing transports, loading and unloading everything the Air Force needed in Europe, North Africa, and Turkey. In time, all of the orders were spit out on IBM computer punch cards. In one three-year period, CHAD received about 168,000 computer orders. Even during the early years of CHAD you could find about 400,000 spare parts from mach-meters to oxygen masks. On most days, you could find up to thirty-two different types of planes, ranging in size from the behemoth C-130 Globemaster to F-100 jet fighters; from the B-50 bomber to the F-100 Super Sabre.

If you thought it was just one big military candy store, it would only be half the story. Anyone who could legitimately press his nose against the window for a peek at the goodies stood a good chance of having his sweet tooth satisfied.

Retired Air Force officer Jerry McAuliffe, in his book, *The USAF in France 1950 – 1967*, pointed out that it offered something for everybody. CHAD acquired parts by the thousands, hired private contractors, maintained air force equipment, and met the Cold War-driven responsibility of procuring new aircraft from NATO partners. An example would be the F-86K jet fighter, which began showing up after assembly by Fiat in Turin, Italy. In fact, Fiat could easily be described as the "poster child" for the evolving Air Force procurement program. This was uncharted territory for hundreds of pilots ordered out of their cockpits and pushed behind desks to confront stacks of paperwork, generally in triplicate.

The idea was simple. With guidelines set in Washington, the U.S. military was to do its part to rebuild the economies of Western Europe, and it made no difference if the countries being helped were former allies or enemies. Uncle Sam would provide the money and expertise so that hundreds of thousands of unemployed workers throughout Europe could put cash in their pockets while contributing to NATO's Cold War efforts. Unfair or not, the U.S. government and military found it easier to deal with the big guys like Fiat regardless of where they stood during World War II.

Among the military hardware produced by Fiat for Mussolini's Fascist Army were armored cars that rolled into action against the Russians in the East, and the British and Americans in North Africa. The thirty-two year old scion of the Fiat family, Gianni Agnelli, was wounded twice while commanding one of his family's armored cars on the Russian Front. In true flamboyant Italian style, Agnelli, while serving in North Africa, was shot in the arm by a supposed German ally in a bar fight over a woman. Agnelli recovered from his minor "love wound," eventually to control more than three percent of Italy's work force, almost 4.5 percent of the country's GNP, and more than 16 percent of its industrial investment in research.

Economic chaos reigned in Europe during the early 1950s. Fiat, like other European industrial giants, faced tough times. Although many firms were rebuilding, what would they make and who would buy it? NATO, and particularly the U.S. military, provided the answers. So what if along the way it provided industrial moguls like Agnelli with the impulse income to purchase their water-borne toys. Thousands of Fiat workers had their jobs thanks in large part to Uncle Sam.

The American Cold War commitment to NATO rested on a foundation of irony that included a house of mirrors capable of disorienting while it beguiled. What was real? What was illusion? Fiat and the Agnelli family, and the citizens of Berry strode through NATO's house of mirrors at the same time. One with a sense of entitlement, and the other with bewilderment.

Fiat and the Agnelli family never lost sight of who they were during World War II and the Cold War. The aristocratic family might have found the strutting Mussolini disagreeable, but not the profitable contracts he was handing out. It seemed pre-destined that NATO would come along with Uncle Sam's fat pocketbook. There was no need for illusions about the future.

It would be hard to imagine that Gianni Agnelli devoted much thought to Fiat's contract for the F-86K, or to Châteauroux, the ultimate destination for many of them. He was busy becoming Italy's most important and best-known capitalist. Agnelli was his country's most famous native son, a close friend of Henry Kissinger, and long time pal of David Rockefeller. While Fiat cars blanketed Europe, they never caught on in the United States, where Fiat came to mean "Fix It Again, Tony." It hardly mattered to Agnelli; he went on to become an adviser to Chase Manhattan Bank and part-owner of Rockefeller Center in New York City. Such were the guidelines for entitlement and wealth.

There were no such guidelines for the citizens of Berry. Cold War demands had transformed a poor, backward region of central France. People are still trying to figure things out.

CONSCIENTIOUS, LOYAL, AND DEPENDABLE

I interviewed François Bayard in 2005 at his restaurant and bar in downtown Châteauroux. It might have been thirty-eight years since CHAD padlocked its gates, but Bayard conveyed a sense of wonderment over the loss.

"It was de Gaulle who decided the Americans had to leave. When the Americans started to leave, there was a wrecked economy. Châteauroux was a black hole."

Before going into detail, he abruptly arose from the table and strode to the bar where he retrieved a sheet of paper from a file folder. Returning to his seat he proudly pushed the paper across the table for me to read. "You can keep that if you want it for your book," he said.

CHAD OFFICERS' OPEN MESS
United States Air Force
Châteauroux (Indre) France
APO 10

5 September 1962
To Whom It May Concern:
Letter of Recommendation

Monsieur Francis Bayard has been employed in the Châteauroux Officers' Club, Châteauroux, France for the past 2 years as a Waiter and Bartender. He assists in the requisitioning of bar supplies, bar inventories, vending machine accounting and has served American and French dignitaries. He is capable in every respect and performs any duty assigned in a very outstanding manner. He is conscientious, loyal, and dependable in accounting for his business transactions and I would not hesitate to employee this young gentleman when the opportunity might arise.

E. M. Warren Jr.
Major USAF

Bayard and assorted colleagues were in the "hospitality" business—important, to be sure—but they were small players in a big picture. International airlines had always been symbols for vibrant national economies. Josef Stalin and the Soviet Union, as excellent propagandists, placed great importance on symbols, the more powerful the better. Europe's airlines had to be kept aloft and healthy, and CHAD was chosen to help ensure this was done. CHAD had maintenance and repair contracts with KLM, SABENA, Air France, Air Italia, SAS, as well as smaller aeronautical firms throughout Europe.

And it worked. The airlines prospered and built their own maintenance and repair facilities. CHAD, once the premier Cold War repair and maintenance depot for these airlines, slowly and inevitably became a shadow of itself. The military had disappeared, the airlines were gone, but the base remained.

At one time, this airport provided jobs for 15,000 people, including 6,000 French citizens. These workers disappeared almost two generations ago, but their ghosts remained. It was hard for me to visualize the millions of man hours of activities that once droned through these silent structures. A large part of that activity was directed by men who had spent most of their Air Force career in the cockpit of a plane and now found themselves behind desks. Meanwhile, civilian contractors started flowing in from America while Air Force pilots were discovering the value of paper clips and staplers.

One Air Force pilot selected for a desk job in procurement was Major James Pattillo. He joined the Army Air Corps as a flying cadet in 1940 and commanded a B-29 Super Fortress on its bombing missions in the skies over Japan. He was a career pilot in every sense of the word. Any thought that Pattillo had been chosen for his new duties as a result of painstaking study would be misplaced. Instead it would be easier to say that Air Force brass had stepped through a looking glass into an uncharted "land of serendipity."

"They needed a special kind of officer to meet the procurement challenge," Pattillo said. "I had no experience in procurement, but I was a pilot and that was important. During those years [1950s] officers were needed to handle responsibilities on the ground and in the air. We had to be experienced pilots who knew what was needed for support on the ground. This was unfair to the pilots, but necessary. After all they were pilots.

"The process that transitioned us from the cockpit to a desk job sometimes seemed haphazard at best," Pattillo recalled. "In my case it was hardly scientific. There were supply depots in Germany, France, and Italy. All of them required command-level officers to oversee procurement. I was told that I was chosen from a list of candidates. I was chosen from among the candidates and sent to Rome because I was the only one on the list who had an Italian name. I was also told that the officers chosen for the other European assignments got their jobs because of how they spelled their names.

"The assignment process could leave you scratching your head, but there was no mistaking what our mission was going to be. It was defensive. We made a decision to defend Europe against any Soviet threat and if the Communists came across, they would know that American units were supporting our European allies. A lot of Americans would be lost, but the Russians knew we were there and within a few hours the Strategic Air Command bombers would be in the air and attacking. We were the trip wire. That was our mission; we were the trip wire.

"But in order to be effective we needed the necessary combat equipment for the mission. A decision had been made to strengthen the air forces of our allies. In Italy, the air force was pretty beat up by us and the Germans. The North American F-86 jet fighter was nearing the end of its production run in the United States. To help Italy rebuild its air force it was decided to send a fighter in bits and pieces to be assembled by Italians in Italy. This was done at the Fiat factory in Turin. We supplied the technical and support help to get the job done. Many of the Fiat-assembled fighters were shipped to Châteauroux."

Colonel Theodore M. Natt was Pattillo's boss in Rome. When Natt was transferred to take over procurement operations at CHAD, he asked Pattillo to join him. Talk about a culture shock; Pattillo and his family were being transferred from Rome to a town he never knew existed until the transfer was in the works.

"Yes, that's true it was both a professional and cultural shock, but this was the end of an era," Pattillo said. "It's like being a sports car driver. You are a pilot, and like the sports car driver, you still need someone to maintain and repair your machine. As a pilot, you followed orders even if it meant a desk job that you were given, because you knew what was needed to keep those planes in the air. Pilots were and always will be the backbone of the air force. If you ever wanted to be chief of staff, you had to have been a combat pilot."

THE PARTY ORGANIZERS

If there ever was a career air force officer's wife, it was Helen Pattillo. She was a woman well prepared to meet the challenge of Châteauroux after living a life of comparative luxury in Rome. She was eighty-seven years old when I interviewed her in 2007. The couple was married in 1942; two years after James entered the Army as a flight cadet. She had learned how to cope regardless of the circumstances. Châteauroux was not Rome, but there was a constant that she knew she could depend upon, and one that would be waiting for her family when they arrived.

"We arrived by train, and on arrival we were billeted at the Bachelor Officers' Quarters (BOQ). It was like we were expected. The boys played basketball and there were many other activities. Oodles and oodles of people came by as we were settling in. This was typical, but there was no denying there was an enormous cultural difference between Rome and Châteauroux.

"Rome was a city of history. It was all around you everywhere you looked. We had a great number of friends. We were attached to the

American Embassy, and of course, socializing there was wonderful. It was a wonderful experience with many wonderful memories.

"I arrived in Châteauroux with a lovely wardrobe I purchased in Rome. I can't remember shopping in Châteauroux for anything other than toiletries."

Among officers' wives, rank meant a lot. Pattillo's superior officer was Colonel Natt, who was the third-ranking officer on the base. Pattillo was his top assistant, and this meant something. Career officers' wives had to have special social antennae in order to fit in. Helen Pattillo didn't need an antenna; she simply fit in.

"I was entertainment director at the Officers' Club. There was always something going on—dinner parties, dances, bridge, bingo games and even musical skits. One in particular, *Guys and Dolls*, was excellent. We also performed it at the NCO (Non-Commissioned Officers' Club) and they really liked it. We had our little social group of six and we knew how to have fun. We'd shop in Châteauroux's antique shops and visited the mushroom caves. And of course, we had dinner at the Officers' Club."

Officers were pulled from the pilots' ranks without rhyme or reason to fill newly created desk jobs. Major Pattillo, who would retire as a colonel, was among them. In an air force pursuing a hitherto uncharted mission there was a substrata of other unwanted employment opportunities offered to unsmiling pilot candidates, not unlike throwing darts at a dartboard in hopes of hitting the right target.

Captain Marty Whalen knew the process well. He couldn't have been happier piloting a C-118 transport, the Air Force equivalent of the Douglas DC-6. The Air Force was very careful when talking about the C-118s that flew in and out of CHAD, dubbing them "surveillance planes" when "spy planes" would have been more appropriate. The 7120th Airborne Control Squadron at CHAD put three to four C-118s in the air every day, 365 days of the year. Of course, it would be impossible to believe that the French didn't know what these planes were doing, but with Charles de Gaulle having perfected his *grandeur de la France* speech, it would be unwise to flaunt it. Problems at the CHAD

Officers' Club burst to the surface, and Whalen found that his dream job piloting a C-118 spy plane was about to end.

"There was no authorization for an officer to run the club. NCOs were in charge. There had been ongoing problems at the club and the base commander wanted them fixed with an officer in charge.

"The Air Force personnel people sometimes work in strange and mysterious ways. Since there had never been an officer running the club, choosing one was a bureaucratic journey of discovery. They obviously had gone through the files of several officers and found that I was their guy. They wanted someone with experience along that line and apparently it didn't matter how shallow that experience was. Looking through my file in 1962 they found that I had organized a few parties for my squadron in the past. I guess you could say that qualified me for the job. I wasn't very happy about it. The decision wasn't mine. I thought to myself, this certainly isn't a good career move. They were having financial problems. They were losing a lot of money. That was intolerable. A lot of things were walking out the front door and a lot were walking out the back door. There had to be strict accountability for the food and everything else. That was my job."

The Officers' Club offered top of the line booty for either the black market or to meet the personal needs of the civilian employees who worked there. Asked if he could remember any examples of theft, Whalen recalled an incident that was at once humorous because of the ingenuity involved, and sad because of the necessity for the action.

"We had been watching this one French kitchen employee. We had our suspicions that he was stealing stuff. He had been closely observed but nothing was discovered.

"One day when he was searched, we found he had taken a whole chicken, split it in half and attached it to the small of his back by his belt. Ordinarily when you pat a man down it doesn't include the small of his back. There's no telling how long he had been doing this. It was sad in a way because he wasn't stealing anything for the black market, but to feed his family."

Motivation for theft wasn't always so noble. There was a bank of slot machines in the club that could be tempting. "I found it hard to believe that we had left unprotected money in the bar every night for the morning crew to pick up and deliver to the bank. Oh, it was maybe one hundred or one hundred fifty dollars.

"This presented an easy target for a thief. We had a night janitor, a Frenchman, who would take the money to play the club slot machines. When the money had disappeared two or three times, the janitor came under suspicion because he was the only one there at night. Of course, we told him to take a walk. We never did find out whether he was lucky or not at the slots."

KALEIDOSCOPE

CHAD was a kaleidoscope of big and small missions. There was a newness about the whole operation, and as is almost always the case with new endeavors, there were bumps along the road.

In 1951, you couldn't fault the American GIs, contractors and French workers who openly cursed the wisdom of choosing Châteauroux for the base. It was easy for them to visualize faceless top brass in the Pentagon sitting behind their desks with congratulatory smiles on their faces. As far as they were concerned, Berry had it all—a strategic railroad, good weather and predictable winds that provided excellent year-round flying. Good weather? Would it ever stop raining? Could the mud get any deeper? Daily winter rain made it impossible for the ground to absorb any more water and it turned into acres of muck.

The Office of Information Services worked hard to put on a happy face when describing the base's evolution in a 1956 publication, *The History of Central Air Materiel Area, Europe*. But there was no ignoring the effort needed in transforming 1,400 acres of French countryside into an air depot unparalleled in Europe. "The unquenchable good humor and resiliency of the American boy and man was rarely put to so prolonged a test as that given to the inhabitants of Tent City and Quontown over the four years from 1951 to 1954," the publication noted.

The "bumps" during these four years took on mountainous proportions. One can only imagine how, if copies of this publication had ever reached Congress and taxpayers back home, it would have set off a wave of apoplectic seizures. Page after page of photos depicted how a lack of tactical planning and preparedness had led to consequences that would have been laughable if they weren't so serious. Tens of millions of Uncle Sam's dollars were at stake.

Visualize a huge garage sale in the United States—or in France, *une très grande brocante*—that stretches for miles; hundreds of big wooden crates that seem to have been thrown around haphazardly. And they kept coming, with no end in sight. It seemed like the shipments and the bad weather had formed a pact; as long as you keep coming, I'll keep raining. And when it did become possible to move all that stuff under cover in old, leaky French hangars, the chaos continued. One picture showed a shapeless pile of boxes, cans, and other unidentifiable containers that were supposed to be protected. They contained food and other perishables. The caption—in classic understatement—remarked, "They could not be stored in any orderly fashion and a certain amount of spoilage was inevitable."

The writers and photographers assigned to complete the publication noted that they were "much interrupted and often distracted" because they had to work in a "hurly-burly atmosphere, much like trying to live in a house in the process of being built." So, under conditions like this, who can blame the writers for developing a florid narrative style exemplified by descriptions like these? "The condition on the outside made it necessary to work inside. It had one great resulting benefit—packing and storing of perishable commodities against the prolonged action of the worst and most destructive of weather-action became for the Air Force an exact science," wrote one writer, who obviously missed out during the perishables' decaying period. "When the weather was so inclement that outside work was prohibited, tireless hands took up the tremendous task of bringing order out of chaos," rhapsodized another writer in the very best alliteration.

In seemed that after four years, things were falling into place. General Hicks, his mission accomplished, was succeeded by "the big man" who, when he arrived in November 1953, brought a feeling to the base that "spring was in that November air." Colonel Lawrence B. Kelly, who would earn a brigadier general's star at CHAD, seemed predestined for the job. As a combat pilot he led the bombers of his

494th Bombardment Group in the Pacific Theater during World War II in air strikes over Japanese positions. Immediately upon arrival he pledged to improve just about everything: living, working, and recreational facilities. Addressing American GIs and civilian employees, he said, "Let's get with it!" To the French, *"Au travail!"* It was like storming the Bastille. These words became his battle cry, and the men and women under his command got pretty excited about it.

Military and civilian personnel responded to Kelly's exhortation with hard work, which in turn led to better conditions and increased leisure time. Their social lives improved both on the base and in town. There was no better barometer to gauge progress at the base than the arrival of the Red Cross with its fabled "Doughnut Dollies." The GIs might have lusted for the offerings of the "Hostess" saloons, but there had to be balance. That was the job of the Red Cross.

Mariana Foringer and her assistant, Elizabeth Holbrook, arrived in early 1954 only months after Colonel Kelly had taken over as commander of CHAD. A short time later, they were joined by two other career Red Cross workers, Sally Bull and Dorothy Gibson. They rented the first floor of a building owned by a Madame Joulet in downtown Châteauroux. Madame Joulet lived in an apartment upstairs. It started as a "bare bones" operation in a large empty room that opened onto the street. Extensive renovation was required. An exterior photo showed four prim and proper women in Red Cross uniforms, low-heeled walking shoes, and big smiles. When I interviewed Mariana in early 2007 she was living in the Hollywood section of Los Angeles. Talk about a small world; her home was only a few blocks from our stateside home.

"The club started with what we thought were the necessary essentials to make the young airmen at home. There was a small library, tables and chairs for the men to relax. The staff provided several things like soft drinks, coffee and refreshments. Also, card games. We were busy from the very beginning."

When asked if there was acceptance by the French, Mariana described how duty and responsibility have no boundaries. She remembered three hard-working French volunteers, total strangers who had come to the Red Cross door.

"There were three women who had come to offer their services to us and to the American airmen. They were Madame Desjobert, Solange de Longerue, and a younger woman named Helene Genin. I believe that Helene was little more than a teenager. Madame Desjobert was an older woman from an established Châteauroux family. She had done similar work for the military during World War I. I've forgotten her age, but no matter, she had the energy and willingness to help.

"There is one enduring memory I have of Madame Desjobert. She invited the staff and volunteers to her house for a dinner. It was a good-sized house; perhaps you could call it a villa or chateau. It was an older home in an upscale area, but it didn't appear to be well taken care of. It was a nice gesture.

"Throughout my assignment, we tried very hard to integrate with the locals. We went out of our way to be part of the neighborhood. There was one restaurant down the street where we usually ate. It wasn't fancy but the food was good and not expensive. Attempts were always made to speak French.

"This was very important if we were to mix successfully. You have to realize that a lot of the airmen were boys, teenagers away from home for the first time. It wasn't surprising that they would not be comfortable being in a French restaurant. So we would take groups of them out to eat, showing them the proper use of utensils and the proper way to order. We had outings to various places of interest, chateaus for example, picnics out in the countryside. Again, these were mostly young kids away from home for the first time, and in a foreign country where almost none of them spoke the language."

Mariana's stay in Châteauroux wasn't very long. She left to marry an American electronics engineer who had been working at the base and then transferred to Yugoslavia. Hers was another one of those "small" missions, a mission accomplished.

1953 – COLD WAR POTPOURRI – 1953

- The Korean War ended on July 27th, with a "ceasefire" agreement signed at Panmunjom, where nine days later the start of the prisoner of war exchange dubbed "Operation Big Switch" began.

- In December, Hugh Hefner unveiled the first issue of *Playboy* magazine featuring a nude Marilyn Monroe.

- On June 17th, Soviet tanks rolled into East Berlin to brutally suppress a construction workers' strike, which had quickly spread into a city-wide protest against the communist regime.

- The Academy Awards were televised for the first time. *From Here to Eternity* won for best movie. William Holden, for *Stalag 17*, and Audrey Hepburn, for *Roman Holiday* collected the top acting awards.

- General Secretary of the Communist Party of the Soviet Union, Josef Stalin, died March 5th. After a six-month Kremlin power struggle, Nikita Khrushchev succeeded him September 7th.

- Physiologist Ancel Keys created headlines by declaring a link connecting animal fat to heart disease; and Kellogg's introduced Sugar Smacks breakfast cereal with 56% sugar content.

- In May, General Henri Navarre became Indochina commander, warning Paris that "there was no possibility of winning the war." By fortifying Dien Bien Phu, he set the stage for "57 Days of Hell."

CHAPTER 3

RELUCTANT EQUALITY

It is hereby declared to be the policy of the President that there shall be equality of treatment and opportunity for all persons in the armed services without regard to race, color, religion, or national origin.

—President Harry S. Truman Executive Order 9981

President Harry S. Truman's Executive Order 9981 was signed on July 26, 1948, only two and a half years before Major General Joseph H. Hicks arrived with his entourage at the St. Catherine's Hotel in Châteauroux. Truman's order was meant to eradicate racial injustice in the military forever. But the ink was hardly dry when the three Armed Services began their foot-dragging. The Army, Navy, and the newly-minted Air Force created as a separate service on September 18, 1947, hung out the Pentagon's dirty segregation linen for the world to see.

The Marines were against integration; the Navy said it had been working to end segregation since the closing months of World War II; while the Air Force was ahead of the curve. The Army, well, it was in a category all by itself. Secretary of the Army Kenneth Royall told a special Presidential Committee that the Army "was not an instrument for social evolution," and later informed the press that "segregation in the Army must go, but not immediately." It took only one day after the President's order for Army Chief of Staff General Omar N. Bradley to say that desegregation in the Army would come only after it was a reality in the rest of the country.

The statistics were devastating. The Marines admitted in January 1949 that only one of its 8,200 officers was a Negro. The Navy admitted that only five of its 45,000 officers were Negros.

Air Force Secretary Stuart Symington did not procrastinate when he submitted an integration plan that proposed assigning Negros on the basis of merit alone. Unlike the Marines and Army, the Air Force threw its full support behind Defense Secretary Louis Johnson, who proclaimed "qualified Negro personnel shall be assigned to fill any type of position . . . without regard to race."

By 1953, the good citizens of Berry were beginning to see and feel the effects of what was happening at the air base. For GIs, the backbreaking daily workload began to ease somewhat, and they started filtering into Châteauroux on a regular basis. White, black, or Hispanic, they brought with them the heavy baggage that had burdened America for centuries. Some of them didn't even realize it. Among those who did was Sam Herrera, who got up from his knees in a four foot high coal mine near Cortez, Colorado only to find that his search for equality in the Air Force would last for half of his twenty-one year career.

Latinos may have been "Caucasians" to the Pentagon brass in 1954, but it wasn't easy to sell that concept to the lower ranks, as Herrera found out. Sure there was a presidential assurance in 1948 that bigotry in the armed services was doomed, but in the mid-1950s, the long awaited death knell hadn't begun. The unfortunate irony that defined Herrera's Air Force career was based on circumstances over which he had no control. He was a victim of history.

Prejudiced or Ignorant?

There are almost two thousand miles separating the barrios of East Los Angeles from Shaw Air Force Base near Sumter, South Carolina. On May 31, 1943, Los Angeles was convulsed by seven days of violence that became known as either the Zoot Suit Riots or the Pachuco Riots. The Pachucos were young Latinos identified by a tattooed cross in the webbing between their thumb and forefinger. It was a rite of passage to have this distinguishing mark done with a needle imbedded in the end of a wooden matchstick. The typical zoot suiter could be spotted from blocks away with his baggy, radically pegged pants and oversized suit jacket with heavily padded shoulders, all in garish colors. A wide brimmed, saucer-shaped hat completed his anti-establishment get-up.

The riots began when a group of white sailors on shore leave left a bar. Words were exchanged with young Latinos and a fistfight broke out. One sailor was hurt. In response, fifty sailors headed into downtown and East Los Angeles, randomly attacking young Latinos, especially those in zoot suits. Before the riots ended on June 7, thousands of service men joined the attack. Scores of Latinos and nine sailors were arrested; eight of them were released without charges and one had to pay a small fine.

When Herrera walked into the Air Force recruiting office in Cortez, Colorado in 1954, he had turned eighteen that day and knew nothing of the Zoot Suit Riots. But they would permanently impact him less than a year later at Shaw Air Base. He would learn the subtleties of racial prejudice under the guise of the "greater good."

I interviewed Herrera at his home in Albuquerque in 2006. He was about to retire from his Civil Service job that included, among other things, the inspection and testing of Air Force One's jet fuel. He and his wife, Anna, whom he met and married in Châteauroux, lived in a beautifully maintained home in the upscale Northeast Heights section of Albuquerque. The couple's love for gardening could be seen in the meticulously groomed, terraced landscaping that included several fruit trees and abundant grapevines.

The Air Force had made it all possible for Sam and Anna. For Sam, the deep, black coalmine shaft in Southwest Colorado was a dungeon from which he had had to escape. For Anna, it was an ocean and two continents away from the Serbia from which she and her family had escaped. Disillusionment was the unwanted baggage that Sam carried with him on his journey to central France. A teenager's illusions that the Air Force offered a racial equality silver bullet eroded like the drip, drip, dripping of Chinese water torture, as inexorable as it was predictable.

"I had been thinking about the Air Force as a career possibility for a long time. I tried to join when I was seventeen, but my father would not give his approval. So on my eighteenth birthday in 1954 I joined up. My father wouldn't give his approval because he felt that people would think he had done so in order to get rid of me—one less mouth to feed."

Herrera, who turned seventy shortly after the interview, was still trim and energetic. When he retired, he was a chief master sergeant, having been a fuel specialist for the Strategic Air Command (SAC) for most of his Air Force career. Herrera never did get his high school diploma, but passed his General Educational Development Tests (GED) while in the Air Force. After his induction in Denver and completion of his basic training, Herrera was shipped to Shaw Air Force Base.

"When I arrived at the base, I noticed for the first time things that I had never seen before. They had just closed the black swimming pool. I never knew that in the South they had separate swimming pools for the blacks. There were also swimming pools for the officers, non-commissioned officers and an enlisted mans' pool. It was June 1955. I was considered a Caucasian and I could swim in the white swimming pool.

"At that time there was a lot of pressure in the Air Force to stop segregation according to race. As you can see from the different swimming pools, there was still segregation according to rank. The black swimming pool was abandoned. It was still there, but it was empty and not being used. Talk about segregation."

Shaw Air Force Base was not and is not your run-of-the-mill Air Force facility. Originally constructed in 1941 as a major air cadet training center, it would eventually become headquarters for the Ninth Air Force, as it is today. Although in a rural setting, the base and the people who ran it were not immune to the realities of the world outside.

"There was one incident that really opened my eyes and led me to doubt that as a Latino I was equal after all. It happened at the Airmen's Club. Things were segregated—maybe you could call it self-segregation, even though President Truman integrated the military. There was a group of Latino airmen, maybe ten or twelve of us. I was with that group. This guy came into the club and went over to the bar where he was drinking a beer. He looked like another Latino. We didn't know him. But one of our group, a guy who used to be a boxer, went to the bar and said something to him in Spanish and the guy answered him in English. That's when our Latino buddy punched him.

"We thought he was Latino but it turns out he was Italian. Then a bunch of other Anglos came over and a big fight broke out. I didn't get involved. I just sat back in the booth and watched what was going on. The whole thing lasted about five minutes. The authorities looked at the incident as a sort of gang riot. It was something that had to be controlled. They did not want any gangs or gang members in the Air Force.

"I don't think anybody was arrested. I don't think they could prove anything that this was a gang. But it didn't end there. Later I was working in a warehouse when the Air Police came to get me. They said I had to go to police headquarters. They didn't tell me why and I had no idea what was gonna happen.

"All of the Latinos were ordered to report. They didn't ask me, they told me, so I did. They strip-searched me. They had me strip naked to see if I had any tattoos or other marks on my body to indicate if I had belonged to a gang such as the Pachucos. I've never had any tattoos in my life. I had to go through that simply because I was a Latino. I'll always remember this scene as an Air Force racial double standard.

"This incident raised doubts [as to] whether I wanted to remain in the Air Force, but I remembered what it was like crawling on my hands and knees in a coal mine."

THE THIRD STRIPE

Once he arrived at Châteauroux-Déols Air Station, Herrera discovered that three thousand miles of water hadn't washed bigotry away. Air Force Secretary Symington might have had a clear idea of how the Air Force should be run, but others with even a modicum of authority would always have the last word.

"There was a fuel problem, a shortage of diesel fuel at the military housing area called Brassioux. Our diesel trucks serviced the area just like we serviced the aircraft at the base. I was on call so I was sent to Brassioux. One of the houses I serviced was where the head of maintenance, who was my boss, Master Sergeant George Kelly, lived. He was from Texas. He was six-foot-three or four and slender and he did have a potbelly. We called him 'Jelly Belly Kelly.'

"I didn't know that it was his house. I had never been in the area before. I had to top off his diesel tank. He came out while I was doing this and said some very nasty things. It was obvious that he was not happy to see me at his house. He said I didn't need to top off the tank, that I didn't need to be there, that it was obvious I was after his two daughters. I was spying on them. He was adamant. In fact he asked me twice why I was there.

"He was very angry. He said, 'I'll see you on Monday.' When I told him it was my day off he said he didn't care. He was waiting for me. He accused me repeatedly of being at his house only because I wanted to make out with his daughters. His tirade lasted thirty to forty minutes. I think racial prejudice was involved. In the work area on base I heard him several times make derogatory remarks about Mexicans. He obviously didn't like them."

At this time there was a budding romance between Herrera and Anna. We'll hear more about how this romance developed and the other hurdles it faced in later chapters. It could have been an account never rendered if either of them had heeded advice that was so painfully stereotypical, although at times well-meaning.

"After I said I wanted to marry Anna, several of my supervisors tried to discourage me. Even the chaplain discouraged me. He said it wouldn't look good. He said that interracial marriages—that's what they called them at that time—were not good for anybody. He said, 'Consider your kids, what are they going to look like?' Like they were going to turn out to be monsters. Even some of the guys in my section said basically the same thing, that it would be a very bad decision.

"I have no doubt that my decision to marry Anna had an effect on my Airman's Performance Reports (APRs). From that point on I couldn't make any rank. I went into Châteauroux as a two-striper (Airman First Class) and two and a half years later I was still a two-striper. Everyone was looking to be at least a Senior Airman (three stripes) so that when you were transferred the Air Force paid all expenses for you, your family, and belongings back to the States. The APRs meant everything to your career."

Herrera said he never had an APR report lower than seven, with nine being the highest. But when he reviewed his records in 1959, "I discovered that there was only one report. It gave me a 4.2 rating. There was nothing else in my file. They destroyed the good reports. Every airman gets a copy of his reports. I've got copies to prove everything I'm telling you."

Herrera produced three Air Force documents, two of them for the years 1956 and 1957. They were titled "Classification Action Requests" and were basically assessments of Herrera's duties and whether he had performed them according to guidelines established for his job. The 1956 document certifies that he had adequately performed his duties that included all phases of fuel and petroleum operations. These included working with French fuel and petroleum suppliers. The 1957 report indicated that Herrera had performed well enough to be awarded a higher job classification, but there was no mention that this carried with it a recommendation for promotion. There were no Airman Performance Reports (APR's) accompanying these two documents.

"I should have had an APR from Shaw Air Base and others from the first three years at Châteauroux, a minimum of three reports. There could have been more because every time you change the supervisor, there would be an APR. So there could have been as many as five that I'm convinced they pulled from my file. I think the missing reports had been destroyed."

Sam showed me a copy of his 1959 APR that unlike others contained a handwritten 4.2 rating notation in the upper portion of the first page.

"I don't know who had written it. It could have been either one of the two NCOs whose name was on the report, Master Sergeant George R. Kelly Jr., who accused me of stalking his daughters, and Clifton R. Heath.

"It was obvious I had to be transferred from CHAD if I was going to make any rank. I finally got my third stripe at Mountain Home Air Base in Idaho and it took only one review. In fact my supervisor, Staff Sergeant Darwin Martin, pulled me over to the side and asked me why I had not been promoted in all this time. He had observed my work and how well I did it and was curious. I told him that I really didn't know why it took so long. I said there was a clique that ran things in the 3130th Supply Squadron in Châteauroux and I wasn't part of it. Sergeant Martin was younger and a different kind of guy than those in France. I liked him, he liked me, and we got along very well.

"After I got my third stripe at Mountain Home, I never got a rating less than seven and most of them over the years were nines. The two master sergeants in France were older, kind of throwbacks to the old 'brown shoe' Army Air Corps before it became the Air Force, and they were from the South with all its prejudice."

Herrera also learned that race was not the only thing that defined discrimination in the Air Force. "It seemed that almost all of my immediate superiors on the squadron level were Masons. Among my immediate supervisors were sergeants Kelly and Heath. There were several others in my squadron who were also Masons. They were a close-knit group. Once I got an idea of who they were it became very obvious. You could spot them by their rings. They were never shy about flashing them at each other when they got together.

"Without a doubt, being a Mason was key to being promoted. You could see that by how many younger airmen who were Masons moved up the ranks as NCOs. Before I came to Châteauroux I didn't even know what a Mason was. This was natural because there were no Latinos in the Masons.

"In fact, three assignments later, when I was a staff sergeant at Travis Air Base in California, I inquired about joining. The reason was simple. It was a training section with three other NCOs in the same office who were younger, two of them with a higher rank than me. All three of them were Masons. I asked if any of them would sponsor me for membership. Of course, that didn't happen. One of them, Staff Sergeant Clyde Ward, disdainfully shrugged off my request and said, 'All niggers and Catholics are the same.'"

THE GREATER GOOD

Staff Sergeant Ward's racial slur, coming as it did in California, had obviously reinforced Herrera's belief that Freemasonry was a ubiquitous presence in the Air Force of the 1950s, and after all, there seemed to be little effort to keep the affiliation secret when members smilingly flashed their rings at each other as signs of recognition. The exact strength of the Masons at CHAD remains somewhat of a secret. However, to gauge the organization's strength, the elaborate formal installation of officers by the Masons' teenage auxiliary, Order of the Rainbow for Girls, is a good indicator.

The installation of Gayla Weems, a CHAD High School sophomore, as the Worthy Advisor for Arc-En-Ciel Assembly #2 at the base, was worthy of front-page coverage complete with photo by the base newspaper the *Sabre*. Miss Weems, followed by four auxiliary officers, all of them in flowing ball gowns, solemnly strolled past a standing room only crowd of admirers in the Quontown auditorium to receive honorariums. Gayla's father, Captain Joe N. Weems—himself a Freemason—and her mother were among them. The 1964 installation was the eighth to be held at CHAD.

Freemasonry was so feared in the nineteenth century that it spawned national opposition strong enough to field presidential candidate William Wirt, whose Anti-Masonic Party received almost 8 percent of the vote in 1832. By the 1950s, Freemasonry had gained grudging mainstream acceptance, but Herrera was not about to turn the other cheek.

Golden Ghetto is a journey through a house of mirrors, with its distorted perceptions of what was real and what was an illusion. Herrera was a young airman who had distortions thrust upon him, blurring the picture of what he believed would be a mutually beneficial experience when he walked into the Cortez courthouse to enlist. He thought it would be simple to figure out—you did your best and the Air Force reciprocated.

Two years later in Châteauroux, when Herrera revealed his intention to marry Anna, he once again discovered that even the best motive was no insurance that he had any real control. He couldn't even find a priest to support him. Fear motivated the Air Force in South Carolina. Herrera could understand this. But he could not understand why three years of APRs had disappeared for no reason, thwarting any chance for promotion. In each case, his career had been on the line, totally out of his control, and insidious doubts about his Air Force choice took hold. And so did paranoia.

Obviously, as we have seen, Herrera had other simple facts kept from him during his tour of duty in France. Foremost among them would have been to stay away from the home of a tall master sergeant with two adolescent daughters—the same Jelly Belly Kelly who was responsible for signing the APRs vital to the career of a young Latino whom Kelly accused of stalking his daughters.

Kelly's family was among those living in the comfortably insulated Brassioux. There was not a sign in sight to indicate the obvious: there weren't any blacks or Latinos living there. Brassioux was the centerpiece of a squeaky clean environment that brought back pilgrims loaded with rosy memories and racial blinkers.

Most of the white servicemen, civilian contractors and Civil Service workers I interviewed claimed their time in Châteauroux was their best ever. And why not? The media back home had done its best to sanitize what was really happening. Listening to the radio or watching TV during the early 1950s, it would be hard to sense any racial imbalance. There were some tokens. *I Love Lucy* wouldn't be the same without Desi Arnaz as Ricky Ricardo. Where would *The Lone Ranger* be without Jay Silverheels, a full-blooded Indian, shouting "Away Scout!" to get his horse in stride with the masked man's Silver?

And then there was *The Cisco Kid*, which beguiled kids of all ages in America from 1914 to 1956 in silent film, talkies, comic books, radio, and TV. Spanning the decades, it was a romantic tale of the Old West that portrayed the Kid and his constant companion Pancho as the good guys who corrected dastardly deeds and were able to do it in a nonthreatening way. The country loved it. There was an easy acceptance of what was essentially a myth created in 1907 by O'Henry in his short story, *The Caballero's Way*. Each episode ended with a corny, mutually appreciated joke. "Oh, Poncho!" the Kid would laugh. "Oh, Cisco!" Poncho retorted as they rode into the sunset.

In 1951, the Air Force, as a separate branch of the military, had only been in existence for four years, and it was rapidly transforming itself when General Hicks arrived at Châteauroux. Another form of discrimination was already apparent. It wasn't based on the color of your skin, but on what had been the original color of your shoes. Tensions had surfaced for years between holdovers from the old "brown shoe" Army Air Corps and the younger guys who wore the newer Air Force uniform with black shoes.

David Madril had a high school diploma and one year of college under his belt when he reported to CHAD to begin work at the 7373rd Air Force Hospital (AFH). He started in stock control, which familiarized him with the medical supplies, and from there spent a good deal of his tour as a medical technician working in the maternity wards, that is, when he wasn't playing football for the base team.

"It didn't take very long for me to realize the difference between the young GIs like me and the older brown shoes, but you have to remember that a lot of the enlisted men were career guys who went into service during WWII and never did finish their high school education. I advanced in rank quicker than a lot of them because I had skills that they never had a chance to acquire. For instance, I could type, and because of that I had a skill they might have had to hire a French national for. Also, because of their limited education, they couldn't handle language very well. Because of these skills, I was able to advance quicker."

Many of Madril's NCO superiors were "thirty-year men." Many had seen action during WWII and the Korean conflict. Now they found themselves marginalized by a younger breed of enlisted men who could pound typewriters and write comprehensive reports with an ease they had never mastered, despite their master sergeant stripes.

"Philosophically and professionally there was a wide difference. There was a very noticeable resistance to change among the old-timers. These were men who had the rank, but could also see that the younger men had more skills and knew more than they did. This had to be very embarrassing to have men working under you who knew everything, smart alecks. Let's face it, when you were eighteen to nineteen years old, you thought you knew everything. Yes, there was noticeable tension."

Madril's NCO superiors were throwbacks to an era when the U.S. military was looking for bodies first and intellect second. Coping with the demands of an increasingly technical Air Force was creating tensions and misgivings rather than the camaraderie that for centuries had been the glue that held all fighting forces together. It is unlikely that the French, who looked toward CHAD for hope and economic rebirth, were aware that this internal conflict was occurring.

1954 – COLD WAR POTPOURRI – 1954

- On January 1st, General Henri Navarre commander of French forces in Indochina said "I fully expect victory after six or more months of hard fighting."

- On February 23rd, the inoculation of children with anti-polio vaccine developed by Dr. Jonas Salk began in Pittsburgh. Denied the Nobel Prize, Salk was awarded the Congressional Gold Medal.

- On January 14th, two American superstars, movie actress Marilyn Monroe and former Yankee Hall of Fame center fielder Joe DiMaggio, married. Nine months later they filed for divorce.

- On May 17th, U.S. Supreme Court, in Brown vs. the Board of Education, ruled that segregation by color in public schools is a violation of the 14th Amendment of the Constitution.

- *On the Waterfront* won Oscar for Best Picture. Grace Kelly for *The Country Girl* and Marlon Brando for *On the Waterfront* won top acting awards.

- On May 7th, the fortified village of Dien Bien Phu fell to the Communists, ending "57 Days in Hell," leading to an armistice signed in Geneva and the French pullout from Indochina.

CHAPTER 4

MAIN STREET BAGGAGE

Only you can make this world seem right
Only you can make the darkness bright
—Platters' refrain in song "Only You"

Talk about reaching out. In expressing universal yearning, the first two lines of the Platters' 1955 smash single, "Only You," couldn't be any clearer. In their distinctive style, the silky smooth Platters helped erase the line between rhythm and blues—the acknowledged domain of blacks and R&B radio stations—and the white pop music culture.

When you hold my hand I understand the magic that you do
You're my dream come true, my one and only you.

HEARTFELT ACCEPTANCE

In 1956, the Platters appeared in concert at La Martinerie's large recreational center. It was a standing-room-only crowd that included high-ranking officers and their wives, enlisted men with their wives and sweethearts, and teenage kids. Sprinkled among them were a handful of black airmen. This was entertainment pure and simple with racial blinders firmly in place.

There is a wonderful photo taken from the control tower of the former U.S. Air Force base at Déols, now Aeroport Marcel Dassault. It shows a large, smiling throng of former airmen, civilian workers and their families stretched along the tarmac below. I studied the photo with a magnifying glass and was somewhat disturbed to see no black faces. These were the pilgrims who returned in 1998. Almost a decade later, I joined another group of pilgrims at the expedition center in Châteauroux. Later, I examined several group photos that reinforced my earlier conclusion. I approached a man who was among those instrumental in organizing the two reunions and asked him about something that had been puzzling me. I asked Mike Gagné if he could recall any black pilgrims returning in 1998 and 2007. His answer was succinct: no.

The Anglo airmen and their families leaving the Platters concert were certain that they were stepping back into a world of assurances. The mass media convinced them that Ozzie and Harriet's world was there for the taking if they played their cards right. By contrast how many black airmen had anything in common with Nat King Cole? This popular pioneering singer had reached the heights of his profession, and in 1956 NBC gave him a fifteen-minute program, *The Nat King Cole Show*. The next year it was expanded to thirty minutes and a panoply of great artists appeared on the show for scale paychecks in an effort to save it. It took only two years for NBC to realize that sponsors had gone into hiding, and the program was canceled.

In 1951, when the Châteauroux Supply Depot took shape, it had been five years since Jackie Robinson broke professional baseball's color barrier. But he could hardly be described as the common man. Besides baseball, Robinson had been a star running-back for the UCLA football team, and a star performer for the school's track and basketball teams. He had a tumultuous Army career during World War II as a second lieutenant, a two-year stint that included failed attempts to have him court-martialed after he refused to take a back seat in a bus. A chorus of death threats and racial taunts was the price Robinson paid for ending eighty years of major league segregation. It is ironic that during the first years that Robinson and the Brooklyn Dodgers made the circuit of National League ballparks, black soldiers and marines couldn't share the same barracks as whites.

Racial ground breaking in the U.S. military started and sputtered noticeably during the twentieth century. As we have seen, Robinson was never reticent about the racial shortcomings of the U.S. Army. Nevertheless he did have a chance to serve his country.

In 1956, CHAD was not a microcosm of America; it *was* America. The music of the Platters might have blended pop with rhythm and blues, but R&B in popular parlance was still considered race music. Attending the concert was Staff Sergeant William "Willie" Ward, a black air policeman, and his twenty-two year old wife, Madeliene, one of nine children raised in the nearby town of Levroux. The recently married couple met under circumstances that could have led to a lynching in a similar mostly white American town of forty thousand people. Willie, barely twenty, simply "picked her up."

"I had a boyfriend at the time. We just had an argument and I was walking along the street when Willie drove up in a military police truck," Madeliene said.

"I had been with my boyfriend for most of the day and found out that he had dated another girl the night before. Willie had been watching me [giggles] and asked me if I was okay. He could see I was disturbed and asked if he could drive me home. I got in the truck."

Ward admitted that Madeliene was quite an eyeful and worth more than a peek or two.

"I was on duty at the time," Ward said. "I was patrolling the downtown area and couldn't help noticing her. I wanted to stop and talk to her. When I stopped, I saw that she was annoyed and I asked if I could drive her home."

There are two photos of the young couple, one at an outdoor picnic and the second on their wedding day, with the two smiling lovingly at each other.

"I didn't propose to her, she proposed to me. That is, if you can call it a proposal," Ward said. "That's completely true. Just ask Madeliene."

"Yes, we were sitting on a bench in the park at the Chateau," she explained. "I said to him that we are going to get married. I don't think he said a word. He sat there and looked at me."

"When Madeliene's family learned that we were getting married, their reaction was good. I felt I was being accepted. After we were married, Madeliene's father wanted to show the neighbors in Levroux that I was now a member of the family," Ward recalled. "He had this box of pastries, I think they were traditional, and asked me to come with him. We walked around to all the neighbors giving them these traditional pastries while he introduced me to all of them. It was his and his family's way of showing that I had been accepted."

A heartfelt acceptance of a black stranger by a family in Levroux was much different than a mandated acceptance by the military. Not everyone agreed with President Truman in 1948, and during the 1950s it seemed that integration of the military represented a form of force feeding for those with no appetite for it.

"Generally things went smoothly, but not always. There were some incidents, but nothing really serious, except for that one time that I could have gotten into trouble by my reaction. I was working as a desk sergeant one night when an airman was brought in. He obviously had had too much to drink and was causing a problem. When he saw I was in charge, he said, 'I'm not taking orders from a motherfucking nigger!'

"I lost control and punched him in the face. I shouldn't have done it. It was a mistake. Things were worked out and smoothed over, but I could have lost my staff sergeant stripes because of it.

"I found out later that he was from Mississippi, and obviously he had brought his Deep South racial feelings with him."

There was a large number, perhaps a majority, of Southerners among the enlisted men at that time. It was true and probably had been for a long time. There were no jobs in the South that paid anything, so the military looked real good to them.

Madeliene's circle of friends included many who dated black GIs. One photo showed her friends and their black dates enjoying themselves at a house party. Madeliene, born and raised in a typically small and insulated Berrichon market town, realized early on that the sight of a black GI was strange enough to illicit widespread comments. Berry was hardly a racial nirvana.

"Even a prostitute in one Châteauroux bar had nasty things to say about us, but Willie got her to shut up very fast. And my landlady wouldn't allow us to go into the house. She didn't want Willie to go inside because he was black. She did not want a black man in her house. This was before we were married.

"She thought that if she allowed him inside it would hurt her reputation. She knew people would talk about it and she still had to rent rooms."

Major General Joseph H. Hicks, first Commander of CAMAE, first Commander of Chateauroux Air Base, first among airmen and planters.

1.

2.

3.

4.

5.

6.

7.

8.

9.

10.

11.

12.

NOTES TO PHOTOGRAPHS 1–12

1. General Joseph H. Hicks, First Commander of CHAD. He was an Air Transport expert in the twilight of his career when he established his command. Hicks was a gruff spoken commander viewed by many of his men as a martinet who harshly penalized even the smallest infractions.

2. St. Catherine's Hotel downtown Châteauroux where Major General Joseph H. Hicks and his entourage sequestered themselves in February 1951 to finalize plans for the Châteauroux-Déols Air Station. Photo courtesy of Valerie Prôt.

3. Mud and More Mud, 1951–53. The quagmire sucked the boots off workers and swallowed equipment.

4. Tent City in all its muddy glory 1951 to early 1954, ten men in each tent fought for position around a single potbelly stove during winter and sweltered in summer. Photo from Memorial booklet 866th US Army Eng. Aviation Battalion.

5. Enemies Unite. It started when some American kids gave Leandre Boizeau his first bag of popcorn. Here Leandre, a third generation communist, waves to the camera during a picnic with these newfound American friends. It was an unexpected and welcome change for a kid who spent countless hours with his family handing out anti-American propaganda on street corners throughout Berry. Photo from Leandre Boizeau.

6. Downtown Châteauroux, 1951 La Marie (City Hall) on the right, where eleven years earlier German officers alighted from a motorcycle and sidecar to tell City officials they were taking possession of the town. Now the Americans were the new occupiers.

7. CHAD Base Main Road, late 1954. Châteauroux's narrow streets had difficulty handling traffic that included growing numbers of military vehicles and large American cars that came to symbolize Yankee wealth.

8. Late 1950s ubiquitous 2-horse power Deux Cheveaux described by Americans as a tin can on wheels. It was also the symbol of the region's new found middle class made possible by Uncle Sam's deep pockets. It invariably came out second best in collisions with American cars.

9. Communist Party HQ in Châteauroux early 1950s. Well entrenched party workers laid in wait for the Americans. Within six months the party newspaper *La Marseillaise* was spouting out headlines such as: "Châteauroux an occupied city, nightclubs, dancehall and hotels are shady businesses."

10. Anti-American Graffiti. It was everywhere during the first years of CHAD but diminished when communists hid their party cards and decided Uncle Sam's paychecks weren't that bad after all.

11. Hostess Bar membership cards and U.S. military script. For a buck or two, young GIs purchased a sense of belonging and a cozy seat next to their favorite hostess, often posing as a college student earning tuition. U.S. Military script was openly traded on the black market at these bars. Photo courtesy of Sam Herrera.

12. *La Grenouillère* (Frog Pond) membership bar. One of the favorite GI watering holes that dotted downtown Châteauroux. This one came complete with two rooms upstairs that were made available to GI patrons.

Listening to Willie and Madeliene, it was easy to see that an interesting racial dichotomy existed in Châteauroux. Madeliene's landlady, fearing disapproval, if not outright contempt, from neighbors, would not allow a black American serviceman to cross the threshold of her apartment house. However, she could and did readily accept Madeliene's rent money, but found it socially unacceptable to allow her boyfriend through the door because of his color. The rent money that Madeliene handed over was earned at a Châteauroux dry cleaners operated, as Ward discovered, by a color blind owner.

"Madeliene's boss, who lived in an apartment above his cleaning store, threw a celebration party for us when we got married," Ward said. "It was just a small group at his apartment, but it was a very nice gesture on his part. After that, we had our wedding reception at Le Faison, a very popular hotel across from the railroad station."

Ward served out his four-year enlistment and returned to New York City where he had a variety of civil service jobs before settling down as a New York City cop for twenty-two years. During that time, he and Madeliene lived with Willie's parents in Brooklyn, then bought, in succession, two homes of their own. In 2006, they returned to her childhood home of Levroux. They became instant celebrities with the media, especially Willie, who had no problem expressing his abiding dislike for President George W. Bush and his Iraq policy.

BEATING THE ODDS

In contrast to the Ward story, there was the odyssey of Lawrence Anderson, another former black air police staff sergeant, and Janine Meriot, a rebellious French country maiden who married him despite advice from both the well meaning and the prejudiced. When I contacted them in 2007, Janine explained that her husband was in an advanced stage of Alzheimer's disease. But she agreed to an interview and gave me photos of her and Lawrence taken in Châteauroux during the early stages of their romance. One of them shows a lovely seventeen year old girl in a beautiful party dress, and the other a handsome black man, one hand on his hip, the other leaning on his patrol wagon. He was in complete uniform, right down to the .45

caliber automatic on his hip. Janine would recount, sometimes painfully, a journey of more than forty years. You would never mistake their story for a 1950s television sitcom.

"I was born nearby in a small town, Le Blanc. I came to Châteauroux to live with my sister. I was seventeen. I ran away from home because my dad would not let me do anything. I had been working since the age of fourteen and I was sick of it. He was so strict. So I went to live with my three older sisters. My mom knew about it before I left, but my dad did not.

"We were all looking for jobs. My older sisters knew there were jobs at La Martinerie. So I went there and I got a job working for a black family, the Fonks. I'm sorry but I can't remember their first names. The husband was in the military police, he was Lawrence's boss.

"They had a baby daughter and were looking for a nanny to take care of her. She was a beautiful little girl. I would walk with her and care for her, and that's how I got the job. Sergeant Fonk would have all these guys over to their house on weekends. They played poker. My husband really loved to play poker. That was how I met him. I didn't speak any English at the time, and he spoke only a little French. But anyway, that's how we first met.

"For our first date, he took me to the home of Alvin Spain, who was also a black military policeman, a very good friend, his best friend. At that time, Alvin was living with a French woman in a small village outside Châteauroux. He was a great guy who was one of the few witnesses we had at our city hall wedding. Lawrence and I dated for a long time, maybe a year or two before we got really serious. He was great, really great. He was a gentleman. He was gallant. He would open the door for me when we went into a bar. He would take me to places I had never been before.

"I was still working as a nanny for the Fonks. And Mrs. Fonk would give me advice, tell me not to marry a black man. She said don't get married to a black man and go to the States. It's not good over there with a mixed marriage. You would not be happy. I didn't listen to her. We continued our dating and he took me to some great places I had never seen before, like Paris. It was beautiful, wonderful; I had always wanted to go to Paris. It was a dream of mine. All this time while we were dating, I was living with my sisters in an apartment in downtown Châteauroux.

"Lawrence and I were getting quite serious. I didn't care what neighbors thought. If they looked at me and Lawrence funny, I didn't care.

"While this was going on, my older sister decided that my mother and father should meet 'Angi.' By the way, I hadn't mentioned it before; we nicknamed Lawrence 'Angi.' On our first visit, my mom was there, but my father wasn't. He was working, and we would visit my mom when she was home alone and my dad was away at work.

"The reason my father was so angry and against me marrying a GI was because he didn't want me to leave, to go to the States. He didn't want us to leave the country, so I had to wait until I was twenty-one years old so I could get married without his permission. My father ordered my mother and sisters not to attend the marriage ceremony at City Hall. My sister, Yolanda, ignored him, but my other two sisters and mother obeyed. Alvin Spain was there and the mayor was a witness."

"I was pregnant when we got married. When my daughter, Katie, was three months old, my father told my mother he would like to see us. My husband was afraid of what would happen. But when we arrived, my father welcomed us with a big smile and his arms wide open. He liked my husband. They became very close. That my husband was black was not the problem. My father did not want me to leave France; that was the problem."

As Janine recalled, the joy from that belated, open-armed welcome from her father quickly faded when she confronted the realities of a mixed marriage in the United States. Traveling west in 1961 from Angi's home state of Ohio, she asked herself whether Mrs. Fonk had been right after all.

"When we were in Kansas, we went to a hotel and they wouldn't let us register. We left, and with our two young daughters, we had to sleep in the car. We came up with a plan. Lawrence and the kids would stay in the car and I would go in and register. It worked. Driving through the South we did this, I think maybe three times, because we knew we would not get a room otherwise. This made me very angry. But what the people thought did not bother me. I was not afraid of them.

"Things did not change much when we got to McClellan Air Force Base in California. You could see what the people were thinking. They would not look at you or would look down when they met you. And I would say to myself, 'what's wrong with you people?'"

Like Willie and Madeliene Ward, Janine and Angi beat the odds that were stacked against them. Despite predictions of failure, their marriages prospered for almost half a century. It might not have been utopia, but it was as close as they could get.

There was no way for the Andersons to know it, but in 1961, they were traveling through a country in which almost half of the states still had strict anti-miscegenation laws banning interracial marriages. As late as 1967, there were still sixteen states that steadfastly upheld anti-miscegenation as possibly being ordained by the Lord himself. As one Virginia judge extolled, "Almighty God created the races, white, black, yellow, Malay and red, and he placed them on separate continents. And but for the interference with his arrangement there would be no cause for such marriages. The fact that he separated the races shows that he did not intend for the races to mix." This dubious, other-worldly judicial opinion was overwhelmingly rejected by the U.S. Supreme Court in *Loving v. Commonwealth of Virginia* in 1967, a case that sounded the death knell for all the remaining anti-miscegenation laws in the country.

It would be hard to explain anti-miscegenation laws back in France, although we have seen racial prejudice in more subtle forms, such as that practiced by Madeliene Ward's landlady in Châteauroux. At CHAD, it was a case of the "invisible man", as described by Thomas Ciborski, a graduate of Brooklyn's St. Francis College with a degree in accounting. He would become CHAD's last paymaster, as the military contingent dwindled to less than a hundred men.

"I can best describe what I mean by 'invisible man' by what happened when I was shipped from the States for duty at Châteauroux. When I came over from the States, there was a black airman, Larry Miller, who also made the trip. I got to know him casually on the flight and on the train from Paris. I never really had a chance to develop a friendship; in fact, after we arrived at La Martinerie he became, for want of another term, an invisible man.

"After we arrived, it was months before I even saw him again. The black enlisted men basically worked on the flight line, fire department, air police, post office, and freight areas.

"The administrative offices such as finance and travel, where I worked, hardly ever saw a black. The perception was that the blacks weren't qualified, because most of us were better educated, with some of us having college degrees."

During the base's halcyon days, a succession of commanders made it clear to local citizens there would be no racial prohibitions at places of business. This was widely understood and practiced, if somewhat reluctantly. Ciborski recalled how one night it became obvious that things had changed during the waning weeks and months of the base's operation. He noted that when the Yankee dollars dwindled, so did racial tolerance.

"I and another white airman, much to my surprise, ran into Larry Miller. We went to a downtown bistro together and ordered drinks. The bartender looked at Larry and said, 'Non, non monsieur,' and sort of wagged his finger in a negative way. That happened more than once. Toward the end of the night, Miller said 'to hell with it,' and we separated, with Miller going off by himself to a black bar. The self-segregation was something you saw everywhere, not only in the bars, but on the base as well. You could see it in the mess hall and especially the airman's club, where the guys were comfortable in their own groups."

As I conducted my interviews and research, it became impossible to separate racial attitudes from geography. It did not matter if the interview was with a Frenchman, an untutored enlisted man, or a highly educated officer. We met Lt. Colonel James Pattillo earlier when he described the crapshoot method of choosing officers for jobs newly created to meet Cold War threats. Pattillo—a Texan and proud of it—who had been a pilot since 1940, found himself in "procurement" simply because he had an Italian surname. Under his command were both airmen and civilian civil service workers, at least one of whom was a black man.

"His name was Lewis. I don't remember his first name. Lewis worked for a civilian contractor. We had a practice called 'local purchase' which means that instead of buying supplies at the depot or commissary, you went out into the local community.

"There were certain things that we couldn't get from the States. It was mandated by Congress that we purchase these items locally. The purpose was to help the local economy as much as possible. Lewis was a contract buyer and he had contact with the French supplier for certain things, and a problem arose.

"I was brought up in the South and had all of the sensibilities of being a Southerner. Lewis was having trouble with his French source, and it might have been because they were trying to pull the wool over our eyes. I supported Lewis.

"At the same time, it occurred to me that here I was, as a Southerner, trusting and relying upon this man with black skin more than I trusted the white-skinned Frenchman. Does that answer your question?"

Self-Segregation

Time after time, I heard the term "self-segregation," as if this deeply imbedded mantra was a natural byproduct that our military carried with it around the world.

Frank Nollette was a precocious kid who graduated from the CHAD American High School in 1957 at the age of sixteen. Raised as an Army brat, he was the son of Francis C. Nollette, who in the mid-1950s was an Army captain at the huge Chinon quartermaster depot. He was also a dorm rat who commuted from his family's military housing in Chatellerault each Sunday and Friday, a trip that required more than three hours by train and school bus. I asked him if there were any racial problems at the high school.

"First you have to consider that the high school was small, there weren't very many blacks attending class. I don't remember any racial issues or overtones at all, but again, it depended on who you were and where you came from.

"But at the same time, I have little doubt that race could play a part in whether you got a promotion or not. When President Truman's order came down in 1948, we were in Japan. At that time, the only blacks you saw were in the quartermaster or stevedore companies, loading and unloading docks, driving trucks or working the mess hall. After integration began, we suddenly had a black first sergeant in our company. We had never even seen a black sergeant before, and as a boy I had never heard that there were any. But he had time in grade so he got the job. For me, the way I was raised, I had no problem with this. For those from the South, it was a big problem, but it was a problem of their own making.

"For the first time, white guys from the South found themselves nose to nose with Negroes. They didn't expect this. They came from a region where the 'separate but equal' practice was in place. Now they found that everyone was equal and not separated. That was pretty hard to cope with. For many from the South, the military, in this case the Air Force, was a means to get out of a relatively poor region of the country. They didn't expect it to be integrated. When I joined the Air Force, it was still there, more subtle, of course, but still there. The clubs were still self-segregated."

Nollette, as a teenager, may not have discerned any racial problems at the CHAD High School, but in many respects, that was not the real world for the enlisted men shipped to Châteauroux, especially in the early years of the base. Gene Dellinger was among the earliest arrivals in 1951, a teenager from Newton, North Carolina, a town where "separate but equal" did not define a philosophy so much as a strict way of life. He recalled his first night at La Martinerie: "We were given a tent with a potbellied stove. We had an older guy, a buck sergeant, from the old Air Force (brown shoe). He was fabulous. He was very funny, kept us laughing. He was a black guy.

"But that's not to say there weren't racial problems. This was during that critical racial period in the U.S. Once you were in the military, you hoped it would set things right, that things were like they were supposed to be. We were in the midst of racial tensions and the people around me in the tent had to make the best of the situation. You had to do it because it made life easier that way. The black person who was with us in the tent was just like us, no difference and he, like us, had a job to do.

"That black sergeant, I don't remember his name, set up his bunk right next to mine, and of course, he made sure it was the closest to the potbelly stove. He was always joking, kept us in good humor, but he was also very serious, too. He had a girlfriend in Paris. He really loved Paris, and he would go there every weekend that he could. Her name was Suzanne. He was always telling us about Suzanne. She was a white woman.

"The black GI was always looking for young, white, blonde women. There were no black women around."

Tom Ciborski's interview describing his evening of bar hopping with Larry Miller supports Nollette's contention. But it was my inability to get another interview that possibly said as much about the racial attitudes in Châteauroux during the CHAD years. I was told by my researchers that there was an elderly French woman living somewhere in Châteauroux who had defied the racial norms more than a half-century earlier. I met this secretive woman calling herself Irene Smith only once and very fleetingly. This was made possible because of the untiring efforts of Valerie Prôt, one of my two chief researchers and interpreters for this book. The circumstances leading to the incident, and the incident itself, are as follows.

It was the first week of October 2008, when Valerie and I approached the front door of Irene's home on Route D'Argenton, in a middle class neighborhood of Châteauroux. After some hard digging, we had learned that Irene, a native of Lorraine in the northeast of France, had been living there for forty years. I had been prepped by e-mails from Valerie as to what to expect when I arrived from the States, or better yet, what not to expect.

First, the expected, which had been described to me in great detail by Valerie, whose emails are quoted verbatim. "You can't miss the house in the area because it's the only one with a red gate, red shutters, a red door, and red wooden beams. The first noticeable thing, apart from this unexpected color in the neighborhood, is that the gate is locked with a chain almost all day long. There is no name on the gate, and no doorbell to let her know she has visitors." So there grew an impression in my mind of a weird house outlined in crimson, located on a busy thoroughfare that intentionally and impressively shut itself off from passersby. But not always, as Valerie discovered on one of her frequent attempts to contact the occupant.

"Incredibly enough, sometimes around six p.m., the gate is unlocked and completely open. There is a nice front garden and the threshold has a cover over it. You can see small wooden plaques on which messages are written such as 'Talking Helps You to Forget' or 'You Feel Free Without a Husband.' These wooden plaques were very fashionable in the 1960s and 70s, and families would hang these messages, but most of the time inside the house. Oddly, there are two bells on the door, but there had been none outside on the gate. One bears the name 'SMITH,' and the other one, 'Irene Smith' which is not a local (French) name. A small sticker advises you to be insistent when ringing, because Irene can be in the backyard gardening. And that's true, you do have to be insistent even if, when she opens the door, she doesn't really look like a gardener.

"But I couldn't see much because she only partially opened a window which was part of the door. It was enough to notice that she was extremely beautiful and classy. She wore a perfect white blouse and I could see the belt of her quality black skirt. She had jewels and a red 'stripe' on her hair. Her hair was combed just as if she had spent the afternoon at the hairdresser's. I could smell her perfume through the little window. I'm sure she put a drop on before opening. She also had a blue beauty mark on her right cheek, and was perfectly made-up as if she had been expecting me or another visitor."

Valerie said she was sorry to bother her, and then explained that she was there to seek a possible interview with Irene for the author of a book about CHAD, for whom she was working as an interpreter. Then the tables seemed to turn, and as Valerie described it, Irene became the inquisitor.

"She asked me many questions about you, about the reason why you were writing a book. I told her that you had never been in Châteauroux at the time of the base, so that she wouldn't think a former boyfriend was back in Le Berry. I gave her the names of people you had met at interviews, of course, she knew most of them. The unexpected question was when she asked me if you were black or white. She told me she had been married to a black man, but had no kids. She said that 'white U.S. soldiers brought racism into Châteauroux because at the time, locals were not racist.' She added that there was a bar on Rue de la Gare where whites would go, and when they saw her they would call her 'nig lov.' She said she couldn't remember exactly, and I couldn't refrain from saying 'nigger lover?' I felt ridiculous and awkward about asking," Valerie recalled. "But she said, '*oui, c'est ça*: nigger lover.'"

When I accompanied Valerie to Irene's home, I found it exactly as she had described it to me. We rang the bell. We waited a few minutes before she opened the door's window. Everything Valerie had described about her beauty was true. Although she appeared to be wearing pajamas, Irene exuded a subtle form of class common to French women. She accepted my business card on which Valerie wrote her local number, but she refused to give me her phone number, strongly implying that she would have to check me out further. She never addressed me directly, but with the penetrating gaze of an inquisitor, Irene checked me out from head to foot. There was no outward sign of what Irene was thinking. She said that eventually she, or someone else, would call me at *La Cure*. The call never came. And why should it?

I had interrupted what had been a successful forty-year attempt at closure, shutting out what had apparently been a painful period in her life. But if this had been the case, why had she kept the name "Smith," and continued to live in a middle class neighborhood of a provincial town in a rural area? Without an interview, this will remain a mystery

1955 – COLD WAR POTPOURRI – 1955

- *The Davy Crockett* television miniseries with its hit song "The Ballad of Davy Crockett" touched off a nationwide craze for 'coonskin caps.

- On December 1st, the historic Montgomery Bus Boycott began in response to the arrest of Rosa Parks, a black woman who refused to give up her seat and move to the back of a city bus.

- In January, the first direct U.S. military aid began arriving in Saigon to support South Vietnam and in July North Vietnamese Communist leader Ho Chi Minh went to Moscow to secure military aid.

- On September 30th, young movie star James Dean was killed in a head-on auto accident in Cholome, California. Only 24, he had completed only three motion pictures, but a legend was born.

- Walt Disney's world expanded on two fronts: the first Disneyland opened in Anaheim, California on July 12th, and the Mouseketeers made their first national television appearance on October 3rd.

- DuPont introduces Dacron, an artificial no-iron clothing fabric that made millions of housewives across America very happy.

- On May 9th, West Germany joins the North Atlantic Treaty Organization (NATO) and the Soviets counter on May 14th with the creation of the Warsaw Pact signed by eight countries.

CHAPTER 5

YOU NAME IT, WE HAVE IT

The hound is restricted from areas marked 'fievre aphteuse' to
discourage the spreading of hoof and mouth disease.
—Restrictive Warning to CHAD Hunt Club Members

How could the Châteauroux-Déols military complex be considered a true Golden Ghetto if it did not offer a program that included activities for even the least athletic? The Air Force decided that it couldn't, and as a result offered the effete snob, and the muscle-bound and fleet-footed alike a long list of sports to choose from, and race was not a factor. Blacks, who composed barely ten percent of CHAD' military personnel, shed the "invisible man" description and rose to prominence when they donned their team uniforms, boxing trunks, and gloves. As it was in the real world back in the States, sports could be a great equalizer.

In 1951, the chaotic conditions at CHAD precluded giving any serious consideration to fun and games when GIs were living in tents heated by potbellied stoves, showering under fifty-five gallon fuel drums with showerheads dripping water, and having their boots sucked off by mud as they waded their way to the latrine.

Pilot Jack Warren, who was among the first C-47 pilots assigned to La Martinerie, recalled that in 1951, the bizarre was commonplace. For example, in one case, putting together the ingredients for even a simple game became an exercise in futility. Warren recalled "one shipment I remember that made no sense at all. We got three thousand ping pong balls in a huge crate, but not a single paddle," Warren said. "It was mass confusion, even to the point where Jeeps would come up missing and nobody knew where they were."

Three years later, the misplaced ping pong balls faded into memory and a sports megalopolis worthy of the emerging Golden Ghetto began to take shape. On July 9, 1954, work began on three recreational areas, including tennis, volleyball, handball, badminton, and basket-ball courts. Over the next twenty months, a skeet range, a ten-lane bowling alley, a Little League baseball program, another tennis court complex, and a horseback riding club helped keep adults busy when not working, and the kids out of trouble.

An examination of how CHAD military and civilian personnel enjoyed their off-duty hours could easily be described as a study of the improbable. Let's start with the "Hunt Club." Doubtless there were few Americans who would have imagined themselves "riding to the hounds" and yelling "tally ho" across the flat landscape of Berry.

A photo of the club members mounted on their steeds could have been taken deep inside Virginia's hunt country or on an English rolling estate. There were smiles all around on the faces of the men and women in their riding livery. But they could not ride hell-bent for leather wherever they wanted. The base newspaper warned the hunters, under instructions from the U.S. Department of Agriculture, about hoof and mouth disease. It also cautioned restraint. "All fields and forests are open to huntsmen except those marked by local land-owners *chasse garde.*"

Fencing has often been described as "physical chess" because it demands mental as well as physical discipline in order to control the intricate foot and hand movements. At CHAD, it was a gift to those who felt they did not fit in with any of the mainstream sports. Fencing opened the door to an elite sport where lunges, parries, and thrusts allowed even those disenchanted with organized team athletics to puff their chests with pride. Fencing Master Phillipe Membre was imported from Paris to teach. One can only imagine Membre's bewilderment when he met his new charges.

Warrant Officer William Smith organized and supervised all of the CHAD adult sports program including the base teams that competed throughout Europe. As was the case with all of the teams, it was up to him to find a place for the fencers to do combat. "Ninety-nine percent of the matches were against the French, who took their fencing very seriously," Smith said. "Phillipe [Membre] would contact the fencing clubs in France to arrange for the matches, which were held at private fencing clubs all over France. These were elite, members only clubs for aristocrats. We had a lot of wide-eyed fencers, two of them just off the farm. Transportation to these matches was supplied by the base."

If you were looking for the typical CHAD fencer, it would undoubtedly be René Coté, whom we met earlier.

"The fencing club was a great thing for me," said Coté, a farm boy from Vermont. "I didn't like other sports, couldn't care less. I really liked the epee and got quite good at it."

Organized athletics at the base could be transformative, while keeping airmen and dependent children at the base occupied and out of trouble. For Dave Bankert, a student at the CHAD High School and son of Lt. Colonel Ward Bankert, playing games eliminated any chance of racial myopia. As the son of a ranking Air Force officer, Bankert lived in a world of class privilege. Bankert described with pride how his father's friendly bets with a black, non-commissioned officer allowed the family to cross the rigid Air Force class lines and subtle racial barriers. It was all about touchdowns and steak dinners.

"Remember, this was 1956 and things were a lot different then. I played football and my best friend was Bob Edwards, the star of the team. I was white and Bob was black. My dad and Bob's father had a deal that for every touchdown I scored, Bob's father would buy the steaks, and for every touchdown Bob scored, my dad would buy the steaks. Bob is dead now. He was really a good player, big and strong. Bob—and I'm not kidding—could throw the football almost seventy yards. We were a small school and played six-man football. I played end, and all I had to do was run down the field under Bob's passes, catch the ball and run in for a touchdown.

"Socially, we ignored most of the Air Force's accepted and seldom-violated rules of rank and privilege. I went to his house many, many times. Like I said, he was my best friend. There were rigid class boundaries in the Air Force, but my father wasn't like that. When my father bought the steaks, he brought them to the Edwards' house for barbecuing. Or it was the other way around when the steaks were brought to our house, and we all got together."

Bankert played every sport at the CHAD High School, and was part of a jock subculture not unlike those you would see in any mid-size town back in the States. Special privileges accrued to the Big Man on Campus (BMOC) even at small high schools. Bankert learned and appreciated what it meant to be the highest scorer on the school's basketball team.

"I got into a little bit of a ruckus with our French teacher, who was French. This girl that I was going out with was in the same class. I don't know how you want to write this, but she had very large breasts. The teacher made a snide comment about that and this infuriated me. After all, she was my girlfriend, I was taking her out and a statement like that was not called for.

"I decided to knock him on his ass. I jumped out of my seat and went up and challenged him. He reported me to the principal, Cliff Gunderson. Gunderson suspended me. I was a star on the basketball and football teams, so some heat was put on him and his suspensions didn't last very long. I continued to play."

JIVE TALK

David Madril had earlier self-consciously described how the older, "brown shoe" holdovers from the Army Air Corps were marginalized because they lacked skills that were increasingly necessary in the new Air Force. He laughed when I described the six-man football played by the CHAD High School team during the 1950s. David was sixty-six when I interviewed him at the 2007 CHAD reunion in Châteauroux. He had a day job at the 7373rd AFH, but football was his first love.

"I played football every one of my four years at CHAD. When I played football, it was on a team that was mostly black. When we went to Berlin and other cities for games, we all went out and socialized together. But when we got back to Châteauroux, we hung out with our own groups, our friends from the same squadron. In effect, we were self-segregated.

"In town I can say that I never saw anyone who would not serve a black man. I think it was more a cultural thing, being with people you felt more comfortable with.

"But sometimes you had to accept some of these cultural differences, even get good at them, in order to be accepted. I remember that back then I even spoke differently. So-called 'jive talk' was becoming popular. Playing on a football team that was mostly black I knew I had better learn how to talk and understand jive talk in order to communicate and get along with my teammates. I spoke jive talk to the extent that some of my white friends had trouble understanding me."

The so-called "self-segregation" at CHAD or in the hostess bars mirrored what could be found in almost any multiracial community in the United States and, as in the States, sports was a door opener.

In 1956, David Bankert's junior year at CHAD High School, there was an old, derelict French Air Force hangar that was transformed early each morning into a basketball battleground. Until Bankert showed up, it was exclusively the domain of twenty-one black airmen, and he was not welcomed. Bankert explained that he had to earn his way onto the court.

"This old French hangar had been converted into a basketball court. I was the only white player who showed up at six-thirty every morning to play three-on-three basketball. A wooden basketball floor had been installed three feet above the original hangar floor."

Three-on-three basketball, also called "three-man pick-up," is intense "smash mouth" basketball; there is a lot at stake. The first team to score eleven baskets is the winner, but the team had to win by two baskets or it went into overtime. That's when it sometimes got rough— when two equally talented three-man teams pushed, elbowed, and grabbed in order to intimidate the other side. There could be as many as ten three-man teams waiting on the sidelines for their chance to take the court.

"In order to play, you had to put down a quarter," Bankert said. "It didn't make any difference if I had shown up early and my quarter was the first one in what became a line of quarters, they didn't let me play. They said I was only one player and I needed three. So there was no chance of me getting onto the court. The twenty-one black airmen made up seven complete teams, and there was no chance that any of them would play with me. But I was a sixteen-year-old kid and I wasn't about to give up.

"There were two players who I'll remember as long as I live. Their names were Jim Rasberry and Chuck Taylor. They won every day, and as you know, as long as you keep winning you keep the court. If you lose, you might as well go home the wait is so long. But one day, they lost and went to the back of the line. My quarter was next in line to take the court, so I asked Rasberry and Taylor to round out my team.

"At first they were somewhat reluctant, but they accepted my offer. From that point on we never lost. I was totally accepted because I could play their game. They nicknamed me 'Shotgun.' Rasberry, Taylor, and fifteen to twenty other black players would come to all my high school games. They would all sit together in the same place in the stands to cheer me on, shouting 'go, Shotgun, go!'

"I'll never forget those two guys. They accepted me for who I was, and they brought along all their friends to cheer me on and support me."

There was no swimming pool at CHAD, but the list of the base's sports activities would leave you breathless if you recited them without a break. There was something for everyone; boxing, bowling, judo, soccer, softball, wrestling and tennis. Considering that baseball extended out of a passel of Little League teams to the base team that toured Europe, that the basketball team competed in age groups starting at age eight, and that the other teams competed from the intramural to squadron level at CHAD and as base teams competing throughout Europe, one sees how a magnificent sports wonderland was created. Among the chess-playing officers class, there was squash, skeet shooting, and turkey shoots for men and women of the Rod and Gun Club.

PROTECTED MANHOOD

In putting together the base teams, Warrant Officer William Smith was handicapped by the fact that, with rare exceptions—usually because of oversights—the best athletes from the states were cherry-picked by his bosses before he had a chance to get his hands on them. "The two sports powers in Europe were SHAPE (Supreme Headquarters Allied

Powers Europe) and Wiesbaden, which were much bigger than CHAD. Being headquarter outfits, they could look at the records of the personnel coming from the States and would try to take the top picks of the players for their teams. And they did," Smith recalled.

Despite built-in obstacles like this, a sports monolith began to take shape at CHAD, and every so often, a top college athlete slipped past the screeners in Belgium and Germany. "We had one player named Pat Duff who played football at USC; another guy by the name of Clarence 'Chick' Fry who played football for Maryland and also coached for us; Butch Hollander a Rutgers man also played," Smith said. "There was also Bob O'Neill, a football player from Notre Dame, who went onto a pro career with the Pittsburg Steelers."

Basketball players were also cherry-picked by the coaches at SHAPE and Wiesbaden. Because the Wiesbaden championship team had to honor a prior commitment, it opened the door for runner-up CHAD, that with the help of a few borrowed players, came within one point of walking away with the European International championship.

Smith described a set of extraordinary circumstances which had been put in place before the CHAD base team was invited to compete for the *Coupe de la Méditerranée Latine*—essentially the European championship. It was an international basketball tournament in Antibes-Juan-les-Pins, an upscale beach resort on the Mediterranean between Nice and Cannes. "To start, the Wiesbaden Air Base team should have received the bid but had overbooked its schedule, and once we got the invitation we plucked Dean Smith and Forest Hanson, a Harvard man, from an Air Force base in Germany to help fill out our squad."

Eventually it was realized that the tournament's motivation to invite the CHAD team had little to do with its on-court prowess, but rather was due to the man who roomed with Hanson when they were undergraduates at Harvard. Geneva-born Aga Khan attended schools in Nairobi, Kenya, and Switzerland before graduating from Harvard in 1959 with a B.A. Honors Degree in Islamic history. Now drop in the names of legendary international playboy Aly Khan, and his wife, movie sex-goddess Rita Hayworth, and you realize this was not the usual tourney invite.

"As far as I know, this is the one and only time that an American military team was invited to a tournament like this, and I think it had to do with Hanson's association with Aga Khan," Coach Smith said. "They put us up in a great hotel at the beach. I can't remember the name of the hotel, but it was big, it was white, and it was right on the Mediterranean. We were given a French interpreter immediately when we arrived. She noted that Aly Khan's son, Aga Khan, was Hanson's roommate at Harvard. My guess is that our team was invited to this tournament because of the association with Aga Khan.

"So after we arrived, we were invited to meet the family at their mansion. It was very impressive and is still probably the only forty-five-room house I was ever in," Bill said. "You walked through the front door and there was half of the leg of an elephant made into an umbrella stand. Walking up the stairs to the second floor, there was a landing with an entire stuffed pigmy elephant. Dean Smith's first wife and I were sitting on this davenport. She reached over to the large coffee table in front of us and picked up this yardstick, and it was made entirely of platinum. It was a very nice lunch, complete with champagne and everything else you would expect. We ate, thanked our hosts, and went off. All of us, the entire fifteen members of the team, were there."

Dean Smith, who played for the University of Kansas basketball team before entering the Air Force as an ROTC lieutenant, had helped the Jayhawks win an NCAA National Title in 1952 and earned runner-up status in 1953. He had probably never in his lifetime competed in a tournament such as *Coupe de la Méditerranée Latine*.

It did not take long for the tournament at Antibes-Juan-les-Pins to get under way, and it became obvious that Smith was the "go to guy." "He played guard and he was an exceptional basketball player," the coach said. "If you needed someone to sink a basket under pressure, then you wanted to get the ball to Dean Smith. We were in the finals against the Spanish national team. To get there, we beat the French national team and the Italian national team. We were leading by one point in the closing seconds, and they had the ball bringing it up court.

The Spanish guy dribbling the ball came directly at Dean, and it was obvious to everyone in the house that he wasn't going to stop. Dean covered his eyes with one arm and his groin with the other because he knew the guy would try to run right over him. And he did. Believe it or not, they called the foul on Dean Smith. They made both of their free throws and they beat us for the championship by one point.

"It was the French people and officials who really protested. They wanted our team to lodge an official protest, which would cost twenty-five dollars. The French put up the money for the protest, and of course, the protest was overruled," Smith said.

Dean Smith survived the frontal attack at Antibes-Juan-les-Pins and went on to earn NCAA Hall of Fame honors as coach of a North Carolina Tarheels team that in thirty-six years won 879 games, two National titles, and appeared in eleven Final Fours.

THE GENERAL'S PET PROJECT

John Natt and his older brother, Ted, were stars of the CHAD High School baseball team. As a high school student at CHAD in the late 1950s, it was important for him to be "cool," and playing sports was his ticket. "Sport was very important. Our baseball team traveled to other air bases in France, quite a nice experience for a young kid. We played games at bases in Paris, La Rochelle, and Orleans. This was to earn a chance to go to Germany and compete for the European championship. Sports were a big part of our life at the base."

Surprisingly, golf was the great equalizer. A quintessentially elite sport in the States, golf at CHAD was democratic and devoid of class boundaries. It was the brainchild of Major General George Ferrow Smith, who was described to me as a golf nut. Smith viewed every bare and flat field in and around CHAD as a potential fairway. If there would be no restrictions on who could play on the course, then by golly, Smith mandated that both officers and enlisted men would help build it. John Natt recalled that during the early stages of the nine-hole course's construction, Saturdays were dreaded.

"That was the day that everyone was ordered by Smith to work on the course during its early stages. And it didn't make any difference who you were or what your rank was. My father, Colonel Theodore M. Natt, was Director of Procurement and the third-ranking officer at the base, and he still had to go out there with everyone else to work on the course. Stooping shoulder to shoulder, they moved up and down the fairways picking up rocks and stones. I believe this also included the driving range. And of course, everyone hated it."

The golf course offered equal opportunities for everyone. General Smith placed great importance on how his pet project was completed, but, once completed, the military's tried and true job placement process took shape. Think back for a moment how career pilot Lt. Colonel James Pattillo was given an important procurement desk job in Rome because he had an Italian last name, and how Captain Marty Whalen, another pilot who had never had a desk job, was given no choice when picked to straighten out CHAD's troubled Officers' Club because he had organized a couple of small squadron parties. Enter Master Sergeant William Clyde Alger, who, with no experience, was selected to manage the golf course.

Alger's daughter, Judy Mesmer, was eighteen when her father was transferred to CHAD in 1958. "He simply saw the job posting on the board, and thought it was interesting. He was a career communications expert, and as much as I can remember, had never played golf. In fact, I think he had never been on a golf course before. He held the job for three years, the entire time our family was in Châteauroux."

LIKE LIVING IN LAS VEGAS

From the day she arrived at CHAD, Barbara Bush realized that she had walked through the gates of a Golden Ghetto. Barbara, who lived there from 1961 to 1966, said, "I must say, it was the best five years of my life. I had never experienced anything like it before, coming from a small town. The social changes—it was so dramatic. It was party, party, party. It was like living in Las Vegas. There were things going on all the time, new people coming through all the time, especially officers who were TDY [temporary duty]. These young officers were always looking for a good time."

Barbara was born and raised in Mohawk, a small town in upper New York State, where she returned after her retirement. In essence, she had never spiritually left CHAD, and realized that to relive her five years in the promised land, it would have to be done with youthful vigor. Barbara was one of the chief organizers of the first CHAD reunion in 1998.

"Before being assigned to Châteauroux I was working as a civil servant on a base in Utica, New York, and I heard how wonderful it was working in France. So I applied, and my paperwork went through."

Barbara's mundane job as a Civil Service Secretary merely offered her something to do during the day. For her, life began at sundown.

"I lived at the Bachelor Women's Quarters (BWQ). It was a great set-up. We were centrally located very close to the Officers' Club. It was easy crawling distance from the club to my room. There was so much to do on the base, and there were always new officers coming through. New guys were a lot of fun. They had new funny stories to tell. It was just a great atmosphere. We went into Châteauroux often to shop at the market and the little stores, and, of course, to the restaurants. I really didn't develop any local French friendships. I barely spoke French, just enough to order water at the restaurants.

"When word came that CHAD was closing, there was just sadness, everyone was sad that this good thing was ending. It didn't affect me as much since my five-year tour was coming to an end. It was just shameful that this wonderful way of life was ending, falling by the wayside and probably nothing like it would ever be seen again."

Listening to Bush, one would be tempted to ask whether Uncle Sam was footing the bill for a global Club Med or a timeshare in Las Vegas. Casual scanning of the base newspaper, *The Sabre*, provided assurance that all was well within the CHAD community. Page after page of stories informed you of religious services, appointment of the new American Legion commander, CHAD's oldest school teacher, and a precautionary MS report. A highlighted date book precluded any guesswork when it came to base activities. Clubs ran the gamut from sports car, chess, jazz, and bridge. For the less sedentary, there was square dancing, ping pong, and ski tours. The offerings at the two base movie theatres were guaranteed to keep testosterone levels low, offering *Pollyanna*, *The Story of Ruth*, *Say One for Me*, and *Tarzan the Magnificent*.

How could any observer find anything wrong with this? It is an overworked phrase to be sure, but Arthur Auster had "seen it all." Auster was thirty-seven when he started work as a Civil Service management analyst at CHAD in 1957, fulfilling a lifelong dream that he would someday live in France. He had come to know and love France through history books, French lessons and a French pen pal. After high school in New Jersey and Washington, D.C., Auster served in WWII and the Korean War with the Navy, and then there were Army and Air Force tours until his military retirement. He ended a long Civil Service career with six years at CHAD.

"My job at the Air Station was to attend staff briefings. I had a staff of from eight to ten people, American military and French. I and my staff had to be aware of all facets of what was going on at the base. This included an assessment of all aircraft, all personnel, the status of the barracks, the fuel and electricity. In other words, the base commander could turn to his command book and ascertain the exact status of what was going on."

Auster was a confirmed Francophile when he arrived at CHAD, and his plan mission was to have his wife and three children follow suit. This meant turning the family's back on the Golden Ghetto in favor of complete immersion in what he saw as the virtues of France. He was finding idyllic romance wherever he looked, but he acknowledged that not everyone shared his romantic perceptions.

"I have heard the phrase that the Golden Ghetto represented, 'sublime isolation.' I talked to people who felt like that, but it was complete anathema to me. These are the people who would say, 'When is my time coming up so I can get the hell out of here?' This was not the experience I sought.

"There is no question that CHAD ranked at the very top. I agree with those who say that the marriage of the Truman Doctrine and the Marshall Plan saved Europe. The American money that uplifted the Châteauroux economy was an example.

"I and my family benefited enormously because it was a 'we thing' and not a 'my thing.' Our three children grew up speaking French, attending French schools and never attending American schools. So we have a strong attachment to the friends we still have there.

"I realize that Châteauroux was a backwater city, but it was my city. I was living in a village of small houses with dirt and stone floors and with a pot boiling over a fireplace with the only food. This was extremely appealing to me. I would say that I loved all of our neighbors. Our children did. Our daughters visited those people at their homes. So it was not economic or social status, it was just that they were fine human beings. They were very earthy people. I worked so hard to get the people on the base out into the community to meet the people. I organized French youth camps so they could interact with French children their same age."

The French acknowledged his work with three commendations, and in 1963 he was nominated for the U.S. Army and Air Force's Ambassadors Award for his work furthering relations between the two countries.

"It all had to come to an end. In 1963, I decided it was time for the family to go back to the States. My three kids had begun to think they were French."

If there was ever an unwitting poster child for the American kids who sequestered themselves in the CHAD Golden Ghetto, you need not look any further than Peter Nyberg. Nyberg graduated from CHAD High School in 1961. Along with his brother, he was a standout football player for the high school team until a near fatal motorcycle accident in 1959 caused massive head and brain injuries.

Nyberg was one of the high school kids Auster had not been able to drag outside the warm and sheltering Golden Ghetto and into the surrounding countryside.

"Do I have any regrets? Now, very much so. I'm older and I guess you could say a little more grown up. I really miss the fact that I hadn't taken advantage. There's so much I missed. My wife and I come over now and we really like France, the way of life. It's more relaxed. I'd like to come back here to retire. Getting involved with the French at that young age would have been good, and I regret that it didn't happen.

"It would have been nice if it had been emphasized with us kids how rich this area was historically and culturally. In fact, I didn't learn until this trip back [that] we were in the middle of [French novelist] George Sand country. We took a walking tour of Châteauroux and we saw things that we didn't even know existed before. For three or four years I was here, and I didn't know anything about the town except La Martinerie and the bars.

"If I could relive my three or four years here as a kid, the first thing I would do is learn the language, and learn more about the city and who the people were."

One wonders whether the Service Club, operated by Kay Wood and supervised by Bill Smith, offered even a minimal list of George Sand biographical books or pamphlets. Sand (born Amandine Lucile Dupin), a feminist who refused to admit it despite appeals to join the cause, is considered the first French female novelist to gain a major reputation. Though born in Paris, she was raised for most of her childhood by her grandmother in Nohant, no more than a two-hour drive through the countryside from Châteauroux. She was to use the setting for many of her novels while caring for a fragile Frederic Chopin, her on-and-off lover. In effect, Sand put Berry on the literary map, a map that literally and philosophically was hard to find at CHAD.

Auster returned to Châteauroux as one of the two chief stateside organizers of the 1998 reunion. The other was Barbara Bush. This fact alone spoke volumes as to what CHAD was all about: two people sharing the same experience, with totally different viewpoints as to why that experience represented probably the most memorable time of their lives.

1956 – COLD WAR POTPOURRI – 1956

- On October 24th, 20,000 Soviet troops moved into Hungary to crush a city-wide student protest in Budapest. Their arrival sparked nation-wide anti-government rioting that took three months to suppress.

- The federal minimum wage rose to one dollar and the after-tax income of the average American rose $63 from 1955 to $1,700 giving a family of three a weekly income of $74.04.

- Ray Kroc opened his first McDonald's hamburger stand in suburban Chicago, inspired by the low cost, kid-friendly fast food served by Dick and Mac McDonald at their nine franchises in California.

- The Platters, who performed later in 1956 at CHAD, were among the music headliners in the first Rock 'n' Roll movie, *Rock Around the Clock* which opened March 21st and starred Bill Haley and the Comets.

- On December 2nd, Fidel Castro and a small group of followers landed in Cuba taking refuge in the Sierra Maestra Mountains to prepare for the overthrow of dictator Fulgencio Batista.

- Republicans Dwight Eisenhower and Richard Nixon were overwhelmingly re-elected, with Nixon extolling during the campaign that "The militant march of Communism has been halted."

- On November 18th, Soviet Premier Nikita Khrushchev told Western ambassadors at a reception at the Polish embassy in Moscow that, "History is on our side. We will bury you!"

CHAPTER 6

CORN, WHEAT, OATS, AND HOMES

The process entailed a unique form of international financing with American farmers providing the initial seed money.
—Government document, declassified in 2001

The Golden Ghetto concept throughout France owed much of its success and reputation to the super efficiency of American farmers. In January 1957, there were probably few wheat farmers in Kansas or corn farmers in Nebraska or Iowa who had an inkling of what Congress was concocting for the sale of their surplus crops. A month later, the appropriate congressional committees and the Bureau of the Budget approved the surplus commodity housing program for France. Documents declassified in January 2001 solved the puzzle: More than 50 million dollars in seed money came from surplus crop sales for the construction of 2,800 military housing units in France. Initially, 507 of these units were planned for construction in a flat, barren field just outside Déols. This number was reduced to the 350 ranch-style homes that became Brassioux.

Culled from the documents was an overview provided by U.S. Military Engineer James S. Arrigona in 1958. "A short time ago, soldiers and airmen were still living in tents or the back rooms of hangars, working under the most difficult of conditions. Today, the tents have disappeared, yielding to buildings, which have literally transformed American military installations all over France. Traffic flows over smooth paved roads, and office buildings, and administrative and social facilities are in service in most places."

Arrigona said Brassioux, because of its size, the layout of its streets, and overall design, "is a unique and most complex project under United States Army Construction Agency, France (USACAF). It is entirely different from any other project under construction in France and one of particular interest to American farmers." How did this unique financing plan work? An international commodities broker received surplus U.S. commodities from the Commodity Credit Corp. and then sold them on the world market, excluding the U.S. and France, for foreign currencies acceptable to the French government. Then, "as construction progresses, the contracting officer authorizes payment to the [French] builder through the medium of Construction Progress Certificates."

Even a reverential Francophile like Arthur Auster, a man who turned his back on the Golden Ghetto, recognized that the 350 homes in Brassioux were a good thing. "It was a town that blossomed then, and it has blossomed even more since. You could call it a gift for the French after we left," he said. And what a gift it was! In 1957, in the adjoining town of Déols, 60 percent of the homes had one or two rooms, 43 percent had no running water, 78 percent had no bathrooms. When the base officially closed in 1967, there was a fire sale, with the French—who had been openly disdainful at first—standing in line. Originally, *Compagnie Immobiliere Marc Rainaut et Cie (CIMR)* was contracted to build 2,700 family units at military bases throughout France. A Paris subsidiary, Marc Rainaut Real Estate Company, was given a contract to purchase land for what was to be 615 units near the Châteauroux-Déols airbase. They ranged in size from two to four bedrooms, and all had two bathrooms—an almost unheard-of luxury anywhere in France in 1958. There were no housing allowances. What would have been allowances went toward utilities and maintenance, and to repay the Commodity Credit Corporation. But the units came complete with shiny American appliances, from refrigerators to modern kitchen stoves, and from centralized heating to washing machines.

If Brassioux was ever to have its Boswell to chronicle its day-to-day existence down to the smallest detail, it could not do better than Yves Bardet, a man who lived in a dream world he began to create as a ten-year-old. It was hard to miss Bardet at the 2007 reunion. The theme of the reunion was the American Southwest, as only the French could reconstruct it. He was standing next to a large display anchored to an easel outside an elaborate recreation of a U.S. military field command post as conceived by the French. Neatly clad in sports coat, tie, sharply pressed trousers, and obviously expensive shoes, he was largely ignored by American pilgrims and their French counterparts. Bardet's display did not have the eye-popping, colorful allure that bogus Southwest Americana offered, but on close examination even the most disinterested passerby would see that it represented his lifetime avocation.

Based on a 1961 directory, Bardet created a schematic, giving the location of every house built in Brassioux, including the street address and the names of the tenants. It was meant to be the durable centerpiece of Bardet's traveling presentation, strong enough to be hauled in and out of the trunk of his Chrysler SUV. Traveling throughout Berry, Bardet was not about to let people forget Brassioux's glory days. The highest ranking officers, including the base commander, had their homes in an interior loop near the center of the village. In descending order, color-coded, were the names of other command officers grouped around the interior circle. Then there were the lower grade officers and the highest ranking, non-commissioned officers scattered about. There was one small section for ranking NATO personnel.

Brassioux was an enclave without fences. The large backyard of one home melded into the yards on either side. This gave American kids newfound freedom rarely found at home. Everything was exactly where it was supposed to be, from the red and blue mailboxes fronting each homes to the corner street signs bearing such comfortable names as "Ohio Drive" or "Colorado Avenue." Objects of fascination were the red fire hydrants, mainstays in every American town but uncommon in France, which were placed strategically throughout Brassioux. There is a photo in Didier Dubant's *Base Américaine de Châteauroux-Déols* that shows an American airman pointing to the water flowing from one of these red curiosities during a test, while three French firemen watch intently. With the ever present Air Police patrolling the streets, safety was taken for granted. If the social norms at Brassioux were taken for granted by its residents, it could also be said that it did not take long for the French contract workers who serviced the community to learn and eventually take for granted the residents' prodigious appetite for spending and waste.

Antoine Price is the son of French parents whose mother eventually divorced his father and married an American airman. With his dual background, Price was in a unique position as a kid to view the American presence at CHAD. Price and his wife, Veronique, have lived the American dream in Southern California, owning two small, French restaurants in San Clemente. It required decades of hard work to get them there. For Veronique, it was a hundred-mile roundtrip commute from Anaheim to San Clemente that began at four o'clock each morning. Among Antoine's jobs during those years was one that included driving a funeral home hearse. Price was ten in 1967 when CHAD was padlocked. Living with his maternal grandmother in Vatan, he attended school and socialized with French kids and their families. "I remember this one kid that I got to know. His family was poor. The father worked for the trash department, and he came around in a truck and picked up our trash. His route included Brassioux. He couldn't believe what the Americans were throwing away. So, being resourceful, he would collect these things, mostly toys, and bring them home for his kids."

Sheila Witherington described the unmatched freedom she and the other kids enjoyed. "We could roam around anywhere and feel quite safe. Our mom would give us a dime and send us to a little store for a loaf of bread. We called it 'the deli.' It was quite safe.

"And then there was the playground. That was our meeting place, and the things we did at that playground defied anything kids do today. For example, we would shimmy up to the top of the swing sets and then walk on this narrow pole across the top. No problem. But if we saw our kids do that today, we would have a fit."

It did not take long for the playground offerings that beckoned kids to Brassioux's park to become superfluous. As family after family moved into the village, swings, jungle gyms, and slides shared backyard space with barbecues and webbed lawn furniture purchased at the CHAD PX and commissary. One would hardly imagine that the parents who spent hours assembling all of this stuff were constructing what would become the foundations for Operation Warm Heart. Gene Dellinger explained:

During 1966, we worked with the base chaplain's office and security office on what was called Operation Warm Heart. This included going to the deserted homes at Brassioux where we picked up all of the backyard children's equipment, sliding boards, swings, and jungle gyms. We went into the homes where we picked up canned food items, good stuff that could be used. There were other things as well, but mostly it was the kids' stuff and food which we stored in a warehouse at La Martinerie. The PX and the Commissary gave us an enormous amount of stuff rather than shipping it out. This included clothes and shoes that were very hard to get in France. At Christmas time, it was given to the orphanages.

Five years before Brassioux's world of plenty was abandoned, the quantity of unwanted, sometimes barely used items had grown to the point where collection trucks began roaming house to house. All of this safety and comfort in a land of plenty included the understanding that rank had extraordinary privileges.

THE HOUSE ON UTAH PLACE

In 1961, Base Commander Major General Harold L. Price and his family had become disenchanted with their digs in a mansion close to La Martinerie, and decided to move to Brassioux. Peter Nyberg still wonders at a job his father, Staff Sergeant John S. Nyberg, was ordered to do by General Price.

"When the General moved to Brassioux, he decided that a three-bedroom, two-bath house was not big enough for him and his family, and that he needed two houses. My father was ordered to build a breezeway between two adjoining houses, creating a six-bedroom, four-bath house. When my father completed the project, creating the largest house in Brassioux, General Price (or his wife) decided they didn't like the interior color. The color had to be 'Desert Tan,' which was impossible to find in France. The only place it could be found was in California.

"So, the General cut orders for my father to fly to California, probably in a huge C-130, and bring back enough 'Desert Tan' to paint the inside of his house. The house is still there, just as my father had put it together. I took photos of it. Yeah, I guess you could say rank has its privileges."

Price's mini-mansion in Brassioux was located at 314 Florida Place. You can't miss it. It's a rambling California ranch-style home with exterior landscaping straight out of *Good Housekeeping*. Yves Bardet confirmed the details supplied by Nyberg. This amounted to being an affirmation from the anointed. The ebullient Bardet described how his childhood curiosity evolved into a lifelong mission.

"I could not understand why there were so many Americans in France more than ten years after the war ended. It was a mystery to me, so my research began. I wanted to find out the true history of Americans in France. I had known that there were Canadians [of NATO forces] in eastern France, but at that time I knew nothing about the Americans.

"I was happy. I love Americans [laughter]. If the Americans hadn't had their own invasion of France at Normandy, I would be speaking German."

Bardet was born in Moulin to teaching parents who were free to travel for three months every year, taking him with them. Bardet's youthful curiosity about Americans was initially aroused by the presence of another U.S. Air Force base that operated outside Moulin from 1951–1958. His jobs for Radio France, first as a sound engineer and also as a reporter, gave Bardet countless opportunities for travel throughout France, including to Châteauroux, where his strange love affair with a place, rather than a person, began in 1987.

"That was the year I came to Châteauroux, first for my work and then for retirement. When I visited Brassioux, I saw the house on Utah Place and said, 'I'd like to have that house,' and I hoped someday I would have a chance to buy it. One night when I was working in Normandy, I researched the web and I saw my house. The one that I wanted was for sale—incredible. I bought the house in 2005. The address has been changed. The new address is 16 Allee des Violettes."

Bardet's house was among those put up for sale by Atlantique, the French firm that owned Brassioux. In 1968, it probably would have cost about 20,000 French francs. "I would say that today it would cost

180,000 euros, and it could go to 200,000 euros," Bardet said. "Even in these hard times, Brassioux is a very desirable place to live."

When Bardet was interviewed in 2007, a currency trader would have to pony up a dollar thirty-five for each euro. So let's figure the math. According to Bardet, who was in a position to know, the price of one of the larger homes in Brassioux could cost as much as 270,000 of Uncle Sam's bucks. All of this made possible because overzealous American farmers in 1957 had their surplus crops put on a world commodity market. *Voilá!* More than 50 million dollars was made available for the construction of 2,800 housing units at twenty U.S. bases in France. And if we want to get really ridiculous about this and try to figure out what 2,800 units would bring on France's housing market at the prices quoted by Bardet, it could bring as much as a nifty 560 million euros, or 756 million dollars. Yesterday's Golden Ghetto had become *le meilleur ami* of French real estate speculators.

"I am retired now," Bardet said. "But I have very little time on my hands because I am very busy. I am Brassioux's *syndic* (managing agent). Brassioux is private property, and as its *syndic*, I have to manage it [a big smile]. It's my life." Not all of the roads to the promised land were as smooth and easy as Bardet's.

A Happy Ending

Mike Gagné, my interpreter, and I had only scribbled notes on how to reach the country home of Maria Reh Beddes and her husband, René in Verneuil, a small village deep in George Sand country near La Chatre. We stopped at the roadside gate of a beautifully landscaped home for clarification, and asked a woman if we were headed in the right direction. "Oh yes, it's just down there," the woman said warmly. Then, in a few words, she unwittingly conveyed how it felt to be a French person of property. "The Beddes' home is just around the corner on the right, you can't miss it. It even has two gates."

Maria Reh Beddes, like Bardet, also smiled a lot during our interview. Seated beside her in the dining room of their immaculate home was her husband, René. A small, energetic woman of seventy-seven, Maria had never been interviewed before and she was smartly dressed for the occasion. It was hard for me to visualize that this tiny woman as the fourteen-year-old girl in 1945 who carried her sister Anna, not yet three years old, on her back across three countries in an escape from Josef Tito's concentration camp in Serbia. From the very first day that Maria and her family arrived in Déols, they had doubts that they had made the right decision. The family was guaranteed there would be farm work waiting for them in Berry, but nothing more could be expected.

Maria never worked at La Martinerie or the Déols air base. However, her husband René did get a job digging ditches for electrical conduit to be laid under the airbase's runways. Or, as he described it, "just another lousy job digging a never-ending ditch by hand." For René Beddes, CHAD represented only a short and thankless laborer's job, but for Maria's little sister Anna, it was quite a life changing story. She was fifteen when she left school in 1956 to take a job as a nanny for Staff Sergeant Sonny Kaniho and his wife, Trudy. The couple introduced Anna to airman Sam Herrera, her husband of almost fifty years when I interviewed her in 2005.

Maria relived her family's odyssey, and how an unexpected stroke of good luck in the form of a French newspaper appeal—perhaps a gift from anti-communist gods—fell into Elizabeth Reh's hands.

"It was while we were staying at a castle [Ernstbrunn] in Austria. It was run by the Communists. We thought it was a trap. The newspaper ad was for contract workers in France. My mother knew a man who was in charge of the personnel working there, and everything worked out. After that we went to Vienna to sign up for work in France, it was for a one year contract. We were among the first to arrive at Déols. I remember the empty wooden barracks, not even a stove. There was no plumbing and no electricity. We lived there for three months. That's all there was—empty barracks."

The barracks were in an area that eventually became part of the Déols air base. Few residents remembered them. Elizabeth Reh, her second husband, and her three daughters lived in the barracks for three months awaiting deliverance, hoping that a French farm family would take them away, give them work and a proper place to live.

"The family that decided we would be good for them was the Bodard family. They were important. They owned a company that made rooves for houses, a brick factory, and two farms. We lived at Verneuil, where my husband was born.

"The house they gave us to live in was a real shack, no running water, no electricity. For me, the Bodards were actually slave drivers! We worked for him for two years, and then I quit and went to work in a bakery."

When interviewed, Maria and René had for decades been living in an idyllic region of France, their home overlooking a beautiful valley to the verdant hills beyond. It was easy to see why George Sands had come to love the region so deeply. Getting there was not easy, and unlike other French families I interviewed, they had gotten there without the help of the Golden Ghetto. In their own way, they were rich—people of property to be envied.

The Least Glittering

As I mentioned earlier in this book, Pierre Pirot hardly fit the image conjured by Cold War years of what a Communist boss should look like. Nonetheless, he had no trouble following party line when it came to how Americans tossed their money around, and this, in his eyes, greatly affected how and where the French lived. In effect, Pirot was defining as well as anyone the composition of the Golden Ghetto that Arthur Auster abhorred and about which Captain Francis C. Nollette warned his son.

"Thousands of Americans came. They lived in the most comfortable houses, in mansions, in castles. As a consequence, there was a surge concerning the rents, which also concerned the locals. I'm not speaking for myself as a Communist, but for everybody, the life of the Americans seemed to be easy. They had a lot of money. There were more and more nightclubs where you could find prostitutes. There were fights at night. It was not always very funny, but that's part of their attitude. We can't say that was true for all of them.

"For example, when I worked in a dairy factory, I ate at the restaurant with Americans who were very nice with us. Generally, their attitude with French people was like the attitude of conquerors. This is something I noticed. It's not only due to the fact that they were Americans, but you notice that everywhere when you have a large number of soldiers. There are always exaggerations, even when they are French. That's what people didn't appreciate."

Pirot's reference to "mansions and castles" was not totally hyperbolic. In the case of Base Commander Major General Harold L. Price, we have seen that rank did indeed have its privileges. Price, with what can be assumed was pressure from his wife, traded the antiquated plumbing and damp, cold rooms of a Châteauroux villa for the comfort of a mini-mansion in Brassioux. Earlier we shared the athletic experiences of Dave Bankert, the son of Lt. Colonel Ward Bankert, a ranking officer at CHAD, who Dave said worked on a highly classified program dealing with ballistic missiles. When compared to Brassioux, the Bankert family's initial living quarters in the 410 military housing area in the Touvent section of Châteauroux could best be described as the least glittering of what Châteauroux's Golden Ghetto had to offer. Dave explained how the family could not get out of there fast enough, and his father did indeed begin a search for a suitable mansion.

The "410" was so named because it represented the number of apartments built to remedy the acute housing crunch in the Châteauroux area. It had caused resentment among Frenchmen such as Pierre Pirot, and little fondness among the Americans who lived there.

For a teenager like Dave, the family's cramped apartment in Touvent symbolized all he found wrong with Châteauroux during the winter of 1956–57. "The 410 housing area was something my mother was not happy with. The apartments were small and cramped, the halls were narrow so that two people couldn't walk through at the same time.

"And that wasn't all. If you lived on the first floor, the noise from the floor above was horrendous. The apartments were cold, wet, and dank. I remember we had rain for at least sixty-nine straight days. Would you believe it—sixty-nine days! After my father decided we should get out of 410 housing, he fell in love with a chateau just outside Déols. It was a big place with three stories. It seemed to me that the ceilings in every room were twenty feet high. It even had a basement. We were really out in the countryside, completely surrounded by grapes, plenty of vineyards."

The 410 accommodations awaiting Helen Pattillo, her husband Major James Pattillo, and their two sons might have been cramped, but they were an improvement over the Bachelor Officers' Quarters. As an ideal officer's wife since 1940, Helen had learned to cope with anything thrown her way by the military, and the family's 410 apartment could be dealt with without so much as breaking stride. Helen concurred with Bankert's opinion of the 410 apartments:

"The hallways were very narrow. You were barely able to turn around in them. The three bedrooms were so small that you could fit only a single bed in each of them. It definitely wasn't Rome. I literally had to stuff my wardrobe into the small closets."

There was a subtle, but very real, network that provided a comfort zone for those who were part of it. The Pattillos immediately fit in among the officers' families at CHAD, due in great measure to Helen's decades of hands-on experience with how the rules worked. As the wife of the second in command of the important procurement program, she was also welcomed into a tight-knit group of six officers' wives. With acceptance like this, it was no surprise the Pattillos found themselves near the top of the long list of officers and highest ranking NCOs awaiting the chance to get one of the cramped, but highly coveted apartments in Touvent. Some of them had been waiting for months. "Housing was almost impossible to find in Châteauroux," Helen said. "There was a long list of people wanting the apartments in Touvent, but we were lucky because we were fourth on a very long list."

Landlords in Châteauroux had their queue. If Americans would wait for months to get a 410 apartment, why not provide them a shortcut while putting some extra cash in their pockets? Often, this meant leaving their neighbors literally out in the cold. Nicole Neveu was eight years old and living with her family in the Le Poinçonnet section of Châteauroux not far from Touvent when the 410 apartment complex was completed.

"It was difficult to find decent housing, especially if you were poor," Nicole said. "We were poor kids, our families were not wealthy. Locals preferred letting their houses to U.S. people because they could charge higher rents. My family couldn't afford to pay this kind of rent, so we lived in one room, and then years later my parents managed to build a house.

"I remember the big difference between us and the Americans, especially the things Americans took for granted. With my friends, I would go to the trash bins to see what the Americans were throwing away. We would pick up fashion catalogues and pin articles and photos on the wall in my bedroom. We would play with shoes that had been thrown away. I remember American ladies who had hair rollers on their heads in the morning, and drove around in big red or yellow cars.

"We had a neighbor who worked for an American family, and she came back home with wall-to-wall carpet and rugs given to her by the people she worked for. We had never seen those before."

Inherently, Châteauroux's Golden Ghetto was built upon irony. Example after example indicated the officers entrusted with making Franco-American relations work more often than not had little to do with the French except in official capacities. Interaction between CHAD officers and ranking civilian officials with townsfolk, with infrequent exceptions, was largely limited to ordering from restaurant menus, visiting mushroom caves, and bargain hunting at Châteauroux's shops. Actual socializing was largely with the town's elite, and as we will see later, it could have embarrassing circumstances.

Touvent was at the lower end of the Golden Ghetto's housing spectrum, and, unlike Brassioux, it was very much a part of France. It was a place where American and French, young and old alike, rubbed shoulders on a daily basis. Cultural differences were quickly recognized and, as with Jean-Claude Prôt, were still savored a half-century later. The Prôt home that I visited for my interview in 2007 in Les Grands-Champs was typically French middle class, and beautifully decorated by Nicole. She and Jean-Claude had both retired after careers in banking. You would never mistake the house for being anything but French, except for the basement—it was *Route 66* revisited, lacking only cardboard cutouts of Martin Milner and George Maharis. Prôt had collected vintage highway mileage markers that traced the route from Chicago, to St. Louis, to Tulsa, Oklahoma City, then deep into the Southwest through towns like Tucumcari, Albuquerque, Gallup, Flagstaff, Kingman, Barstow, and finally to the blue Pacific at Santa Monica. Old license plates hung on the wall. They illustrated the Illinois, Missouri, Oklahoma, and New Mexico of the 1930s and 1940s.

"When I was a kid, it was very easy to like the American kids, just the same as I remember and know them as adults today. The American kids even tried to teach us how to play baseball. After the game was over, the Americans would get together and drink Pepsi, and the French kids like me would go out onto the field and pick mushrooms. They were amazed that we would eat them, and what we didn't eat we would take home to our parents. French and U.S. kids lived in neighboring areas in Les Grands-Champs for us and Touvent for them. I remember in June we would pick up crickets and throw them in the open windows where U.S. girls were sleeping. One day we found panties hanging on a branch in a tree outside one of the windows.

"It seemed to us that all of the Americans were very friendly, including the GIs. I remember we would go to a bar called *L'Imprévu* (The Unexpected). There was a pinball machine, and when they had played enough, they would let us play the last ball. That was generous. Then there was my first piece of chewing gum. There was an American worker who was repairing an electricity pylon. He was at the top and he threw one to me with a big smile on his face."

For the French kids, there was a fixation on cars—big, brightly colored convertibles that grew in size with the passing years. To touch them, or better yet, to be lucky enough to ride in one of these chrome encrusted highway monsters was the dream of every French kid. Jean-Claude recalled that the fantasy ride in a convertible could be a double-edged sword. "Two friends of mine jumped into the back seat of a U.S. convertible. They were looking for coins on the floor that were dropped from the pockets of passengers. The owner arrived and said nothing to them. They were sitting in the back seat without saying a word. The man drove to La Martinerie on the other side of town, and my friends had to walk home. It was a long walk."

For poor kids like Jean-Claude and his French buddies, pennies were very important. You got them whenever and wherever you could. It did not matter how you got them, the floor of an American convertible or, as Jean-Claude recalled, panhandling from GIs' wives shopping at the American commissary in Touvent. "We would wait for ladies pushing shopping carts, the first we ever saw because the French shopped with baskets. We asked them for a few cents. When we had enough we would give twenty-five cents to the next lady to go inside and buy Pepsi-Cola for us."

While French kids were on the prowl for loose coins, Officers' families were always on the prowl for good housing at a good price, perhaps even a villa or a chateau. They recognized a bargain when they saw it. Later, we will consider how Ferdinand de Lesseps, of Suez Canal fame and Panama Canal infamy, posthumously provided a temple of delight for officers and ranking civilian officials lucky enough to penetrate this privileged inner circle. The legend of de Lesseps' former hunting lodge in the wooded village of Meunet-Planches has grown over the years, and was still very much alive five decades after CHAD was padlocked.

1957 – COLD WAR POTPOURRI – 1957

- Frank Sinatra headlined his own musical and dramatic TV show attracting the biggest names in showbiz, and denounced rock and roll as "Sung, played and written for the most part by cretinous goons."

- On October 4th, Soviet Union launched the first earth orbiting satellite, *Sputnik I*, followed a month later by the launch of *Sputnik II* with the dog, Laika, onboard, panicking the U.S. and initiating the Space Race.

- A dark year for New York baseball fans as the New York Giants announced that they are moving to San Francisco and the Brooklyn Dodgers said they were pulling out also and moving to Los Angeles.

- On March 25th, in Rome a treaty establishing the European Common Market was signed by Belgium, France, Italy, Luxembourg, the Netherlands and West Germany.

- President Eisenhower dispatched federal troops to Little Rock, Arkansas to enforce Supreme Court mandated racial desegregation of the city schools, allowing nine black students to enroll in high school.

- Jack Kerouac's novel, *On the Road*, was published by Viking Press, introducing "beatnik" and "beat" into the American counterculture lexicon and inspiring the "hippie" movement a decade later.

CHAPTER 7

ESCAPING, EGGS, AND BETRAYAL

My mother was more frightened by the Resistance than by the Germans.
When the Maquis arrived, they would take everything we had.

—Monique Madrolle, Levroux farmer's daughter

What were the good citizens of Berry thinking when the Americans arrived in 1951? Doubtless their attitudes were shaped by the more than a decade's worth of hardship that included five years of Nazi occupation and shared privation following WWII. One military occupation had ended six years earlier, and no matter how you sugar-coated it, another was about to begin. The Golden Ghetto had not yet taken shape. Some Berrichon hoped for the better, while cynics doubted any military occupation could be fortuitous. These attitudes lay in wait for the Americans. A good memory could be transcendent.

The Madrolle family was tentative, not knowing whether they should greet the man who had emerged from the car in their driveway or wait to see who he was. The man looked familiar, but they would have to wait until he got closer in order to identify him. Hans, for his part, was anxious, and more than a little apprehensive as he stepped from his car. It had been ten years since that night in 1945 when he had escaped from the farm outside the town of Levroux during the night and made his way home to Stuttgart, Germany. With him were his wife and young children. Hans had returned to repay a debt which, although small, had burdened him for a decade. The college-trained engineer remembered well how kindly he had been treated by the Ma-drolle family during the six months that, as a German prisoner of war, he had been conscripted as a farm laborer. Hans was returning to show the family that he was not a thief.

Monique Madrolle, who after a long romance would marry former CHAD airman René Coté, was only five when Hans was dropped off by French authorities at her family's farm. The family had learned that

German P.O.W.s held in Berry were being offered to farmers to help with their crops. What they didn't realize is that by accepting Hans they had opened the door to months of surveillance by the Communist *Maquis*. Although she was young, Monique had fond memories of the German prisoner who was dropped in their midst.

"I don't remember his full name. But I do remember that he was a kind person. He had reddish hair. He never spoke bad about the French. My father didn't like to call him Hans, so he gave him a surname. He called him Arthur.

"A German prisoner living on the farm created problems. The prisoner was not supposed to eat at the same table with us or sleep in the same house. There were some men, the Communist *Maquis*, who called themselves the guardians of the German prisoners, who said 'we have to make sure that you don't do them too many favors.'

"The Maquis never gave us any warning they were coming. It was usually in the evening and in the morning. They wanted to know where the Germans were working and what they were doing, which lands they were working, and so forth. I really don't know how many Germans were working on the farms. But I think there could be one working at most of the farms in Berry. Hans was sent to us because at that time we needed help on the farm, and all we had to do was sign up at the office where they registered the prisoners.

"The *Maquis* banged on a lot of farmhouse doors because they wanted to know that the Germans were not treated too well. Not that we were treating them like a member of the family, as we would a normal French worker. We had French workers; these workers ate at the table with us. Hans did not eat with us. But we had to give him hospitality. He slept in a building next to our house that belonged to us. He left on his own. Actually, he escaped during the night. When he left, he took a dozen eggs with him. So we woke up and my father went to tell him it was time to go to work, but there was no answer. He was gone.

"It was a nice surprise when he came back to see us. I believe it was about ten years later. He told my mother that he was returning to pay them back for the eggs he took with him when he left. He laughed, so did my mother and father. He had lunch with my mother and father and left the next day."

When Major General Joseph H. Hicks decided the Air Force should set down roots in Berry, he chose a landscape that seven years earlier was only a few kilometers from a parachute drop zone hotly contested by Communists and Resistance fighters. I found in my interviews that an understanding of this sometimes violent intramural jousting had to be taken into consideration in order to understand the entrenched attitudes Americans faced.

Before starting on this book, I had swallowed, in their entirety, the cinematic hors d'oeuvres that depicted the French Resistance symbolized by the *Maquis*. Nazi resistance was not only courageous, but could be transformative, as we were shown in films like *To Have and Have Not* and *Casablanca*. Most of the attitudes confronted by the Americans in 1951 had been shaped during the preceding decade.

Ironically, the American military could not have found a better testing ground for democracy than the Berry region of L'Indre County, where more than a quarter of voting age French were Communists. The result was sensible, face-to-face confrontation, with each side looking for a "bottom line" solution that might offend, but tried not to embarrass. All that "U.S. Go Home" and "Yankee Go Home" graffiti was meant to be offensive, but in the end, only served as fodder for GI wisecracks. Americans might have been accused of using Uncle Sam's money to commandeer homes and apartments, but landlords had no problem going with the flow by raising rents on their properties.

It was an unfair fight from the beginning. Perhaps the expectations of French workers were unrealistic. Who thought about the future when menial work resulted in paychecks bigger than your poor, working-class father had ever seen? If you were lucky enough to have any English language fluency at all and could throw in some basic clerical skills, an even brighter future beckoned. The Communists tried, but in the end, found no effective way to disparage the Americans, so they threw in the towel and became engaged.

Gerard Charbonnier, our nearby neighbor in Ste. Colombe, also realized that painful memories were buried under the rich soil of

Berry. Charbonnier was a teenager when attempts were made to recruit him for the Communist *Maquis*. "One day, they came to our house and said, 'You have to come with us.' But my parents complained. My father was asked, 'What does your boy do?' My father said, 'He brings in the harvest,' and they [*Maquis*] decided to let me stay home.

"This was a dangerous time. I remember one incident that happened, because when it happened I was with [farm] workers. Eight German soldiers had been killed by the Resistance. We wondered what was happening when we heard the sound of machine guns firing. We all ran into the house. The next day, we found out that Monsieur Renaud, who was a mechanic, had been forced to go with a German officer to the cemetery where they had taken the dead Germans. He was ordered to close the lids of the coffins. This is true. Monsieur Renaud was really scared. I don't think there were any dead Resistance fighters."

The shootings took place in a region that was fertile territory for clandestine Allied parachute drops of supplies for the Resistance fighters. News of these supply drops was widespread throughout Berry, but only by word of mouth, and then only very cautiously.

A Box Full of Treachery

Recruitment by Resistance forces both Communist and on the Right was spurred when it became apparent in 1942 that war-time manpower needs in German factories could no longer be met, and forced labor conscripted from occupied countries would provide a remedy. In France, young men had one of three choices: flee to the countryside and seek out the Resistance; join the pro-Nazi, pro-Vichy *Milice*; or be shipped out to German forced labor camps. Aimé Camus was twenty-six years old when he received a notice on January 21, 1943 from the Vichy *Ministère de L'Intérieur, Etat Français* (Department of the Interior, French State). It read, "You are requisitioned to work in Germany in order to contribute to the relief of French prisoners. You will leave Châteauroux on January the 22nd at 10 p.m. Any default will be liable to severe sanctions."

Camus was ninety-two when interviewed in 2009 at his home at 106 Avenue John Kennedy in Châteauroux. Also seated at the kitchen table was his diminutive and sprightly seventy-nine year old wife, Lillianne, who proffered a covered, stationary-size cardboard box which she said was her "box of souvenirs with the best and worst of memories." As she removed one item after another, it became apparent that it could just have easily been called a "box full of treachery." Spread on the table was an identification tag with his German I.D. number and picture—except for the photo, much like an American dog tag; copies of documents acknowledging his deportation; the original deportation letter; and the unused German passport for his "quick" return to France, dated 6 December 1943. The passport was dated almost five months after he had expected to be home in Châteauroux. Although the passport was late in arriving, Camus believed that it was a ploy that offered hope for the conscripted French workers that they would be allowed to shortly return home when, in fact, the Germans intended to keep them for the duration of the war.

"That evening, I went everywhere people had promised to help me in this case. Actually, I wasted my time. I could have run away and hid somewhere. It was too late for me. Those who had proposed to help me said, 'In six months' time, you will have a leave to come back home for a few days, then the Resistance will be organized and you won't go back to Germany.'

"Those people who made the promise actually worked for the Germans. They told me it was too early, there was nothing to do before six months, and those six months turned into twenty-eight months. It was a catastrophe. I did my best, but I had so little time. And I never came back after six months to be helped by the Resistance. It was the French government who handed us over to the Germans.

"The night we were shipped out, the military made sure everything was under control at the train station. The train was full. There were already people coming from the south. The train station was crowded. In the train, everybody started playing cards. It was maybe a way not to think."

Camus's proficiency as a toolmaker turning out metal parts on the lathe was invaluable for the Germans, who shipped him to a tank factory in Linz, Austria, where he made or refurbished the tools necessary to keep the Panzer Corps operating. When the war ended, Camus brought home with him abhorrence for anything political, and this precluded having any concerted involvement with the Americans at CHAD.

James Besset, a Châteauroux native, was thirteen when he heard some of these stories. Interviewed in January 2009, Besset showed that, even as a kid, he had been aware of the power struggle between the Left and the Right during the final years of WWII.

"At that time, I never heard about precise missions. I only heard about parachutings that took place at night. The pilots were either British, or it was French military who had left France for England to resist. They dropped ammunition, food and money so that the *Maquisards* could survive. There were also messages.

"Most of the time, parachutings took place in regions where there was enough open space and enough room to allow a plane to land if necessary. Fires were lit to allow pilots to see their target or the place where they could land.

"Sometimes the planes brought in men, but also took men out for whom the situation had become extremely dangerous. They were men who were about to be arrested and who had to leave for England."

If these missions were dangerous for the Allied pilots and the Resistance fighters, it could be equally as dangerous for any French who interfered. "I heard about a parachuting in Bommiers [near Déols], in particular, because something weird happened. After the parachuting, the family who lived in the farm nearby was killed. Some people linked the parachuting and the crimes. It was said that the family had kept what they had found for themselves, and that they had probably been executed by the FFI [Right wing *Maquis*]. There were many comments about that story. Nothing was proved."

Léandre Boizeau, whose magazine provided rich archives of materials for *Golden Ghetto*, expanded on the quote that opens this chapter when he recalled how his uncle responded to confiscatory demands by the *Maquis*. "I remember my uncle told them one day, 'Okay, that is the last time. Next time, I take the pitchfork!'"

Maxim Drucet was among those in another group of French workers rounded up not far from La Martinerie by the Germans for deportation as forced labor to Austria and Germany. As his niece, the former Nicole Tuillemain, explained in an August 2009 interview, her uncle had vowed that he would not allow himself to be conscripted. "The Germans came around to round him and other men up to take them to the camps in Germany. He did not want to go, so he ran. And that's when the Germans shot him. They shot him seventeen times. This happened where we lived in La Cité de Vitray. They took my uncle's body to a hangar at the base and just left him there. His wife then went over and claimed the body.

"I'm trying to remember how old he was. Let's see now, he had been married and he had a son and a daughter, so I would say he was about thirty years old. He was a real nice guy. I do remember that it occurred right where we lived. This really caused an outrage, the people were really upset. Everybody was very upset, very mad. I was about five years old when it happened. There was not what you could call a monument, but a memorial at the location where he was shot."

When interviewed Nicole was seventy. She and her husband, former Air Force Master Sergeant, Gerry Lowry, had just celebrated their fiftieth wedding anniversary in Albuquerque.

JUST GOOD FOR BUSINESS?

Pierre Pirot followed in the footsteps of his father and was a life-long Communist. Although he was only fourteen in 1944, one incident that year left an indelible memory. "I remember I went to Maurice Gonon's. He was a Resistant [fighter], and we wanted to know if he could house a Resistant [fighter] for a couple of nights. I was still only fourteen, but I remember a man I didn't know in his house. Maurice Gonon told us the guy was a Resistant [fighter], too. Actually he belonged to the *Milice*, but Gonon didn't know about that. I remember the guy was good-looking, and two of his fingers were missing. Later, we were to find out that he was actually a *Milice* spy.

"The night after, the Gestapo came to Gonon's. He was arrested on April 4, 1944, and was ferociously beaten. He was sent to Buchenwald

[Germany] where he died. The guy who had denounced Gonon was arrested by the *Maquis* near Limoges and he was executed. The *Milice* was harmful whenever they could infiltrate Resistance groups. They did a lot of harm.

"Guys from *Maquis* and guys from *Milice* knew each other. When you know each other, there is less bloodshed, as in Le Blanc for example. The first regiment of France [*Milice Corps*] authorized by Vichy was in Le Blanc. The fact that some people decided to join the *Milice* was not systematically political. It was sometimes for business. For example, Monsieur Chichery who lived in Le Blanc, was a member of Parliament, and was close to [President Pierre] Laval, and became Marshall Philippe Petain's Commerce Secretary for two weeks, and also owned a bicycle factory. He was the one who created ration cards.

"Chichery did his best to have a *Milice Corps* in Le Blanc. At the time of the STO [forced labor], many young guys didn't want to go to Germany, so the possibility they had was first to work in the bicycle factory, or second to join the *Milice* First Regiment in Le Blanc. The situation at the time was extremely complex. In April 1944, France was for Pétain [of the Vichy government]. People changed their minds in June [after the D-Day invasion]."

A sense of fear pervaded Berry during the final two years of the war. A Vichy or Gestapo spy could be anywhere, so it was best to watch what you said and strongly consider to whom you were speaking. James Besset was quite open in his 2009 interview, but one could not help but wonder how he or his parents would have answered similar pointed questions in 1943.

"At that time, we lived in the 'Occupied Zone,' so Germans could overhear conversations. Those ones were easy to spot. The problem was with the *Milice*. These French people worked for the Germans and you never knew if the person you were talking to was for or against the Germans. Besides, *Maquisards* never had a special label on their forehead to let you know who they were. Most of the time they stayed in the countryside. They were hidden in farms or were hired there to be able to act more discreetly. Some lived also in camps in thick woods.

"This uncertainty, who was friend and who was enemy, affected our family. At home, there were no comments, but we knew what was going on globally. And since I was very young, I don't know if adults had private conversations when I was away. People used to say: 'keep your secrets secret' or 'walls have ears.'"

It was no surprise that not a single *Miliciene* could be found for an interview. Without a *Miliciene* in sight, it was difficult to see how young Frenchmen could equate Nazi collaboration during WWII with American fraternization during the 1950s and 1960s. But they did. Lillianne Diez, who dated GIs and eventually married one, was warned by snubbed local suitors that a shaved head awaited her once her American boyfriends departed. She scoffed at this threat, surmising that it was not hatred, but youthful jealousy that prompted these warnings. No evidence could be found that there had been a single shaved head in Châteauroux after the war.

More than fifty years later, it was not hard to imagine that Berry, as part of the Free Zone controlled by the Vichy government, had been relatively untouched by WWII. Driving from Bouges-le-Chateau through Levroux across autoroute A20, veering over to Déols then down to La Martinerie and into Châteauroux, it is hard to imagine that much of this area had been the target of intense and deadly aerial bombings by the Germans in 1940, and the Americans and British in 1944.

Route d'Argenton, one of the main thoroughfares through Châteauroux, had been a favorite target for low flying American and British fighter bombers during the last year of the war. It was wide, and paved strong enough to handle even the heaviest German armor, an excellent escape route for Third Reich forces fleeing north toward home. Royal Air Force (RAF) fighters chopped up and almost totally destroyed a column of Major General Erich Elster's division as it fled along Route d'Argenton. To get home, Nazi forces had to pass over Le Pont De Notz, a bridge where James Besset witnessed scenes of carnage that remained a vivid memory more than five decades later.

"When they [the Germans] left, they were constantly harassed by the Resistance. Day after day, the Germans were finding it harder to get out of Berry. I remember the bombing on Route d'Argenton," Besset said.

"The Germans were on their way to Limoges and the Allies stopped them. One truck was hidden behind the bridge, and one plane managed to swoop and shoot at the truck which exploded under the bridge. There were bodies everywhere. At the time, we didn't have much to eat so even as a kid, I had to do the gardening, and I remember the garden was in the area. I hid in the shed to have the impression that I was protected, and I remember I could see the planes between the boards."

After D-Day, June 6, 1944, low level bombings and strafing attacks by British and American fighter bombers intensified. Châteauroux with its air base at La Martinerie was a prime target. In August, the Allies destroyed the railway station and the gas supply depot, leaving local residents without fuel for two months.

The carnage increased in proportion to the number of Germans fleeing north. On September 7, the Allies twice attacked a retreating German convoy on the road between Châteauroux and Issoudun. An estimated 400 Germans were killed, 300 horses lay dead along the road, and 70 vehicles were destroyed. The next day it became obvious that the retreating Germans, without air cover of their own, were helpless targets in what had become an Allied "turkey shoot." Elster saw the hopelessness of his position and surrendered his 22,000 men and all of their equipment to the Americans on September 12, 1944.

As in all wars, it is often the small atrocity that starkly defines universal insanity. As a young girl, Suzanne Beaujard witnessed one such war-time horror from a hiding place in her home village of Bouges-le-Chateau. Her family lived on a farm in Verdenay only two hundred meters away from Le Fourion, a small stream spanned by what is locally known as the Verdenay bridge. As with the Route d'Argenton bridge, this tiny span, located amid neatly landscaped greenery and seasonal flowers, also hid horrific wartime memories. For Suzanne Beaujard, there would not only be the noise of war, but its bloody image as well.

"The FFI [Right wing *Maquis*] tried to destroy a tiny bridge in Bouges-le-Chateau when the Germans were leaving, but they didn't manage. The Germans fired back. There was a woman near the bridge. I don't know who the woman was. There were people from the north of France who had been sleeping in farms," Suzanne said, and then recalled a vision that indelibly stamped in her mind the true cost of war. "The woman had a baby in her arms and there was blood dripping from the baby's feet. I was a kid and I was hidden in the weeds behind the farm. I think the woman was killed. We were scared to death because we knew they were ready to kill all of us." The Beaujard family's hiding place in the weeds prevented them from being discovered by the Germans, who at that time were holding her father as a prisoner of war in Germany.

Four years before a frightened Suzanne and her family hid in the bushes to escape death, Germany had already identified Déols and Châteauroux as prime targets for the Luftwaffe. La Martinerie, where the Bloch bomber was developed and manufactured, was the ill-fated facility that demanded attention from the Germans and Allies alike. It was hard to identify which side was the enemy. On June 4, 1940, German Heinkel bombers continuously attacked La Martinerie for seven hours, and the following day the Luftwaffe returned to bomb it again along with the French military camp at Pruniers. On June 11, 1940, the airfield was once again bombed, as was the rail station in Châteauroux. Bombings continued throughout the month as the Luftwaffe filled the skies over Berry. At Issoudun, one air strike accounted for 400 casualties, and on July 11th, 44 were killed and 110 wounded in a raid at Levroux.

The Levroux attack occurred three weeks after the Germans officially took charge in Châteauroux. On June 20, 1940, two motorcycles with sidecars parked in front of the city's town hall. Two German officers stepped from the sidecars and met with Châteauroux's mayor, and declared that they had taken over the city. Just short of eleven years later Châteauroux's fate was to be decided again when Brigadier General Joseph H. Hicks and three aides closeted themselves at the St. Catherine's Hotel without so much as a brief stop at city hall. The purpose of the American military mission was officially announced three months later. During that interim, fear and anxiety shared equal footing with hope and anticipation.

1958 – COLD WAR POTPOURRI – 1958

- Modern consumer credit was born, making it easier to "buy now, pay later" when the American Express and Bank of America introduced credit cards to compete with the Diner's Club Card.

- The first television quiz show scandals surfaced with revelations that contestants on *Twenty-One* and *The $64,000 Question* had been coached and given their answers.

- On April 2nd, in response to *Sputnik*, President Eisenhower proposed to Congress the establishment of the National Aeronautics and Space Administration (NASA) to administer scientific exploration of space.

- On March 24th, Elvis Presley reported to his Memphis draft board, and after completing armored training, Private Presley was shipped to West Germany on October 1st to join a tank crew.

- From April 15-17th, the North Atlantic Council met in Paris to discuss how NATO could counteract the Soviet arms buildup in Europe by maintaining a defensive shield of conventional and nuclear weapons.

- The first Grammys music awards were presented, and folk music got hot as the Kingston Trio's hit, "Tom Dooley," rose to number one.

CHAPTER 8

IT AIN'T NECESSARILY SO

In the States, there would be all of the things they wanted but couldn't get in France—washing machines, modern stoves, televisions, modern kitchens and bathrooms. It was the Promised Land.
—Mike Gagné, Châteauroux restaurant-owner's son

George and Ira Gershwin could hardly be expected to know that in 1935, when their opera *Porgy and Bess* opened on October 10 on Broadway at the Alvin Theater, one of its hit songs could easily serve as an anthem for the myriad misconceptions that the French had about CHAD. Delivering the cynical lyrics for the song *It Ain't Necessarily So*, drug dealer, gambler, and all around bad guy, Sportin' Life, advises that De things "dat yo' liable to read in de Bible, It ain't necessarily so."

It was difficult for young French maidens leaving the family farm, and unemployed men looking for work where there were no jobs, to view CHAD as a crapshoot. Sure, there was skepticism, and why shouldn't there be after the Berrichon had endured first German bombings, and then the "friendly fire" from Allied bombers and fighter planes during WWII? Skeptics, even the most virulent Communist naysayers, were ignored when the goodies offered by Uncle Sam were put on display. First, there would be the jobs for men and women alike, and only one disqualification remained constant—no Communists. But as we have seen, even this was only honored in the breach. If your qualifications met the growing demands necessitated by the rapid expansion of CHAD, then you just hid your party card and stepped forward. All for the greater good of NATO. Those first, furtive peeks into the Commissary and PX unveiled a hitherto unimagined materialistic utopia.

Mike Gagné, with dual Franco-American citizenship, was in a unique position to identify the "promised land." The son of Joe Gagné, the iconic owner of the fabled downtown Châteauroux restaurant, Joe's From Maine, might have been a kid at the time, but this did not diminish his awareness. The restaurant, with a tiny downstairs dance hall and bar complete with music, was an instant success, and a magnet for young and lonely American airmen—voilá! A marriage mill in the making. "With the base here in Châteauroux, the women realized all of these things they dreamed about could be theirs if things worked out. A lot of French women's dreams was to marry an American and go back to the States."

It all started in 1952 when a smiling American airman, identified only as Sergeant Martin, posed with his new bride, Maire, and the somewhat somber city official who performed the ceremony at Châteauroux City Hall.

After the Martin/Maire nuptials, things became hazy as to how many marriages were performed in Châteauroux during the years the base was in operation. Valerie Prôt's graduate thesis, *The American Forces in Châteauroux*, later published in book form, places the number of marriages at 452, or 9.6 percent, of all marriages performed in the city during that time. Fernand Marien, widely known as "Gypsy," placed the marriage figure at more than one thousand, noting that there were three thousand young, single American airmen at the base at any given time. "Hostess bars" were Gypsy's specialty. "In 1963, I took over my first bar in Saint Christophe. In 1967, I took over the Crazy Bar, then the Raspoutine, in front of la Caserne Charlier, and then the Lide Rue de l'Amiral Ribourt that I still own, but now it is a brasserie, it is le Kheops." One explanation which adds credibility to a higher marriage total is that not all French country girls were married in Châteauroux, but enticed their boyfriends to their hometowns for the ceremony.

The experience of David Madril, the football playing corpsman who handled supplies for the base hospital's maternity ward, exemplifies to a large extent what the dating scene in Châteauroux was all about. He was nineteen when he arrived at CHAD in 1959, and he had not perfected his timing when it came to women.

"Yeah, I guess you could say in a way Joe's bar's sexual magic really worked for me. My wife-to-be worked at the hospital as a secretary. It was not a quick romance; it had to be nurtured along. That first evening at Joe's was strange. I made dates with two different French women, one of them the secretary at the hospital. There was supposed to be a half-hour between them. But my first date arrived late and my second date arrived early."

The Cave, where the dates took place, was a noisy and cramped basement club below Joe's. A small record player blasted out current rock 'n' roll from the States. A large jukebox in the upstairs main restaurant added to the ambient cacophony. It had a tiny bar and small tables. The mortar of the walls and ceilings were etched with dozens of names. For young GIs, most of them away from home for the first time, and fortified with beer, hot dogs, hamburgers, and ham-and-cheese sandwiches, the Cave was probably the closest thing to a stateside "meat market" in France. It was the ideal place for Madril to hide his naïve duplicity.

"So I had two dates at the Cave, but I had them sitting at separate tables. A friend of mine from the base was there, and I gave him some money to take care of the second lady, hoping that I could switch over later. Looking back, what really saved me was that the Cave was so crowded it was hard to see from one table to another.

"To get my friend into the ballgame, he had to be paid. I gave him maybe five or ten dollars, which was a lot of money at that time. I even paid for a hotel room in town, since she was from out of town and needed a room to stay. I stayed with the first lady and she eventually became my wife. The next morning, my friend Jack told me that he and his date stayed up and talked all night [Laughter]. It turned out that I married the girl I was with, and Jack and Michelle, the other girl, also got married, and were married for forty-seven years. She just passed away."

Madril's marriage ended after nineteen years. He visits his daughter in Châteauroux as often as he can. His two sons live in California.

With American GI money freely flowing from one downtown bar to another, who could blame the French for believing that even teenage airmen were rich. But if they could have heard Madril describe the five or ten dollar bribe he offered to a friend as "a lot of money at that time," it would have raised questions among even the true believers about just how rich these American kids were.

Husband-seeking maidens would have been surprised to learn that the stateside taxpayer underpinnings for the CHAD Golden Ghetto were generated by hardworking, but hardly affluent, families which their GI boyfriends had left behind. Possibly, it was the reason they left home and joined the Air Force in the first place. In 1956, the minimum wage in the United States reached one dollar an hour for the first time, giving a family of three a weekly income of $74.04, and average annual income rose to $17,000. A year later, the average American production worker was making $82.32 a week.

Jeanine Lebeau's appearance belied her sixty-nine years. She was thin and elegant, smartly dressed, with her gray hair stylishly cut. During my interview, her answers were crisp and straightforward, and you had the distinct impression that she did not abide superfluous chit-chat very well, that maladroit questions and replies did not fit her style. Jeanine was twenty-four years old when her son, Laurant, was born. The father was an airman who never saw his son; he returned to the States while Jeanine was pregnant. My interviews with her and her son paint a clear portrait of how the same event can be colored by acceptance and denial. In both cases, pride was a dominant factor—a keen sense of self-worth by the mother, and the unremitting belief by the son that it was only insurmountable circumstances that kept mother, father, and son apart for forty-five years.

Until her meeting with Laurant's father, Jeanine's experience with Americans represented a matter-of-fact way to earn extra money. "I remember working for a lieutenant colonel, his name was Lester Stone, I babysat for him often. He always paid more than he should, and told me to help myself if I was hungry. He had a way to make you feel at home. I wasn't used to that! He also gave me a letter to present to the chief of the Châteauroux Hospital, which greatly influenced and helped me to get hired.

"I'd like to think that Colonel Stone's generosity was typical of the Americans. The Americans were wonderful. They brought us their modern way of living, food, cars, and especially hygiene; they were also very generous. The French at the time were very dirty. The GIs were more generous than the French, and they were also very clean. I belonged to a small group of girlfriends who went out dancing together, but we had nothing to do with the girls who went to 'special' bars to meet GIs.

"One of my girlfriends went out with a GI and she asked him, since I was single, if he had a friend for a blind date. We went out dancing. He was a tall, handsome man, a bit like my son, brown eyes, brown hair. It was love at first sight."

Jeanine referred to Laurant's father only as "Bill," a man whom she would only know for about two months. "I was a bit disappointed by his departure; our affair may have been different if we could have known each other better, but it was too short. Maybe things would have been different then. I didn't know him well enough to see his qualities or his flaws.

"When I told him I was pregnant, he told me that he could not stay with me in France. Anyway, I didn't want to go to the U.S., as my mother was quite ill and I had to stay here to take care of her. I think he may have asked to leave early, or he was forced to leave, because I was just a few months pregnant when he left. I tried to find out if he had tried to 'run off,' but as you know, the Air Force is quite secret on these matters.

"I never tried to contact him, and I don't think Laurant would like to try to find him either. Too many years have gone by, and anyway, we never had any sort of contact with him, he apparently never tried to find us either! On the other hand, I am a bit curious. What I would have liked was to see him next to Laurant to see if there is any resemblance. This was a very short love affair. We didn't even really know each other.

"People were not surprised when they heard of my short affair and that I had become pregnant. Very many local girls were or had been pregnant by GIs. My family did not reject me. They were only worried that I would have to bring up my child alone. I was not the only one, of course. I had friends who also had children out of wedlock.

"It was not until Laurant was around eight years old that I first talked to him about his father being a GI. His first reaction was 'why didn't you leave with him?' I told him that I could not leave his grandmother all by herself; she was too ill. Laurant never held against me the fact that he did not have a father. I think he suffered from this, especially as a teenager, but he never let it show. But the curiosity had to be there.

"When Laurant was younger, people used to say that he looked more like an American than a Frenchman. On certain occasions, when what is happening in the U.S. is in the news, it surfaces. The last time we talked about Laurant's father was on 9/11, wondering if maybe he was there."

There was no resentment in Jeanine's voice, only an obvious sense of pride in what she had accomplished as a single parent, supporting not only herself and her son, but her mother as well. Châteauroux is essentially a small town, and jobs were hard to come by, especially after CHAD closed. Class roles were sharply defined. For Jeanine and her mother, emigrants from Poland, there was undoubtedly the lingering stigma bestowed on those not wearing the livery of native Berrichons.

"After Laurant's birth, I moved three times. I had to find a bigger apartment for us all. I still live in the same apartment for over thirty-five years now. I worked for years for a doctor, and ten years in a lingerie shop, and at night and weekends I would babysit to earn extra money."

Laurant, his wife, and son live on their own farm in Velles, a village of two hundred people not far from Châteauroux. He also works as a landscape specialist for the city of Châteauroux. When I talked with Laurant, it was clear he viewed things differently. While Jeanine accepted that her Bill had simply disappeared, her son had fashioned a lingering perception that his dad, like his mother, was the victim of obstacles that could not be overcome.

"I don't believe that my father tried to evade his responsibilities; my mother did influence his decision as he told her that he could go back to the U.S. and that she would join him later. Things got complicated later on, as my grandmother was quite ill, and Mom decided to stay here to take care of her," Laurant said.

"My mom, being pregnant, was told by a commanding officer that the 'father' had the choice of leaving the Air Force to live with the mother of his child, or to have her join him in the States once the baby was born. I think this situation, my mom having a child with a GI, was not an unhealthy situation. Let me explain: a lot of French girls at the time were pregnant from GIs and tried to hide it from their families, this was not my mom's case. This was because plans had been made so we would all go live in the U.S. and my family knew about this, there was no shame.

"Naturally, I have regrets, one special regret. The thing that comes to my mind most often is the lack of fatherly love and guidance; this is my biggest regret. I have a son of my own now, and my wife is continuously after me because I let him do what he wants all the time. I think I react this way because I want to 'make up' for all the things I missed with my own father."

Toward the end of the interview, I detected an almost wistful longing by a very proud man for something that might have been, despite the lack of evidence to support it. "I truly believe that if things would have been different, I would have grown up in the States as planned by my parents. I believe that my dad must have been a nice person, and I am sorry that I never met him."

A social conundrum was created with the Americanization of Châteauroux, and babies forced local citizens to face a dilemma which, with closer study, might not be a problem at all. Valerie Prôt discovered in her research that being the child of a French mother and GI father, even if he had long ago disappeared, bestowed a certain caché.

"I remember kids who were maybe seven or eight years old living in an orphanage who said their fathers were American," Valerie said. "And they showed me the American stamps on the envelopes from the letters they received. It was fashionable at that time to say your father was American, because it was a dream to them, even if he wasn't there.

"I think with the majority of cases it was true, that the father was American, but in other cases it was not. It also might have been that these kids were lost, abandoned, because they didn't know why they were at the orphanage."

Parents of unmarried, pregnant daughters found that having a grandchild, but no son-in-law, did not necessarily carry stigma. This sanguine parental reaction surprised Valerie. "For the parents, there was shock but no shame that they had a daughter with a child that was left all alone," Valerie said. "But it would be more shocking if she left to go to America and that they probably would never see her again. I think the choice between having a daughter with a baby that had been left behind might be preferable to having a daughter who left to go to America. There was always the hope that a nice French boy would come along and marry her."

Bob Goggin remembers that local parental hopes for their unwed daughter and fatherless grandchild were not universally accepted by a growing number of French women who were reluctant to wait for their local "dreamboat" to come along and rescue them. A number of them came through the Judge Advocate General (JAG) office where Goggin worked as a litigator. "I remember cases where airmen were transferred back to the States, and women would come into my office and say they were either married or had a baby, and demanded to know why they weren't getting any money or government support. There were no easy solutions.

"If the man was no longer in the U.S. military, there wasn't a darn thing we could do. If he were still in the Air Force, there were steps that could be taken to see that some money got back to France to at least support the child. Of course, we felt badly about the situation, but it was out of our control."

Bad situations in the Golden Ghetto were easy to ignore when a young and appealing French woman saw only the end of the American rainbow. Louis Morin had an elevated view of the CHAD dating game. "It was a dream for them [the French women] to live in the States. They did not know what to expect except for what they saw in the movies, and they were hoping to live out a dream. They were trying to find that tall American with blond hair and blue eyes who was going to take them to the States.

"I remember that they were thinking that an ordinary American mechanic was an engineer. A really high ranked engineer and not simply a man who had to work with his hands and a wrench. There was a big difference. There were a lot of girls who would tell their friends and family that, 'I want a big wedding because I am about to marry a wonderful engineer.' Then they would get to the States and be disappointed."

Morin's view was shaped to a great extent by his privileged position on the second rung of France's tiered social class system. With the poor below and the rich elite above, Morin spoke with the secure confidence of a well-established, middle-class bourgeois, and as such, he believed his insight was accurate because it was dispassionate. Very few in his social circle saw the need to escape from troubled families or from poverty into the waiting arms of American GIs. It's quite possible that Morin had fashioned an overview that carried its greatest validity simply because of its vantage point.

When conducting my interviews, I found it necessary to remember that for many of the interviewees, this was the first time they were being asked to relive a life that had long ago disappeared. For some, these memories shaped happy fantasies that were becoming dimmer with each passing year. For others, these memories were reluctantly revisited.

DID YOU EVER HEAR FROM HIM AGAIN?

CHAD was essentially a small town with the same pitfalls you would find along any Main Street in America. There was no guarantee that love and the dream it embraced were not charlatanish. Judy Mesmer was a real "head turner", beautiful enough to bring all action to a halt when she strode her way into the CHAD bowling alley. She was eighteen and already out of high school when her father was transferred to CHAD in 1958, and for three years managed the base golf course. She held several office jobs at the base, and it did not take long for her to get into a social swirl. "There were times when I had three dates in one day. There were plenty of activities at the base, bowling, movies and dancing."

Judy revealed that she had been something of a rebel, a young girl who would take chances, and was very capable of keeping certain aspects of her social life a secret from her parents. She described how she often easily avoided her parents during the evening.

"I went to the Airmen's Club because my parents never went there. But we did go to the Officers' Club if we were dating officers, my parents weren't there either. I dated two officers. There was a big difference in the atmosphere of the two clubs. The Officers' Club was very elaborate. The Airmen's Club was hardly more than a beer hall, but it did serve mixed drinks."

Shortly after she arrived at CHAD, Judy met a GI whom she identified only as Clark, an airman first class. During a two-year period, she said they began to see more and more of each other, but only during the week, hardly ever on weekends. Then Judy discovered she was pregnant, and also discovered that Clark had a secret—one that would change her life forever.

"I'm trying to remember whether it was from his best friend or another friend who knew where he lived. But it was the friend, not the closest friend, who took me to his house, which was not on the base. I don't know what his motivation was. The house was more like a trailer home. Clark came out briefly and we talked. I went along with his friend because he told me that Clark was married and had two kids. The fact that he lived off-base made it possible for him to keep his family a secret from me.

"There really wasn't a 'showdown' that night. He later came to our house and we talked outside. He said everything would be okay. He was telling me that his wife was leaving and other things. I can't remember them all. But I do remember him saying everything will be worked out, we'll write to each other and keep in touch. At the same time he was saying this, he was crying. He visited us a few times. My mom and dad thought he was a nice guy, but they didn't suspect that we had a serious relationship.

"I met with him one more time. It was when I was leaving, when I was two months pregnant. We were in a car and he began to cry. He said he and his wife were separating and after that things between us would be okay. That we would be together. I loved him, and I absolutely believed everything he said, that things would work out and we would be together. But I never heard from him again. Never.

"My father didn't want anything to do with the baby or me. My mother was more supportive, but there was little she could do. So I went back to the States by myself. My grandmother in Rhode Island took me in for several months."

Judy returned to the States basically as a homeless person. During the final seven months of her pregnancy, Judy wandered about, first staying at the home of a friend in Kentucky, followed by a short stay at a Catholic home for unwed mothers. With the delivery date drawing closer, Judy moved back to Rhode Island where her grandmother took her in. She told me that since there was no hope of a family reunion, she reluctantly decided to relinquish her baby to adoption.

"After my son was born, I made up my mind that I would see him the day he was taken away from me. It had all been arranged by a social adoptive service in Rhode Island. I had just left the hospital and the baby had not yet been taken away. It was arranged that I could watch through the window of a five and dime store, I think it was a Newberry's, when my baby was brought out. I guess you could say I had my nose pressed to the dime store's window. A social worker walked out of the hospital with my baby in her arms. My son was wrapped in a blanket. They got into a waiting car, and I watched as they drove slowly away.

"I can't ever remember being that heartbroken in my life. I truly believed I would never see my son again."

Judy's father never accepted her back into the family. After his death, there was reconciliation with her mother. Judy made many inquiries over the next forty years, hoping against hope that she would at least discover that her son had enjoyed a happy and healthy childhood that continued into manhood. Then one day, she discovered her son had been living with his adopted family only blocks away in the neighborhood where Judy's family had set down roots decades earlier in the same Rhode Island city. It was a miracle that turned bittersweet. The reunion took place not long before my interview with Judy in 2007. There have been a few meetings since then, but the fifteen hundred miles separating mother and son did not make things easier, and the relationship has continued to be cool. Judy was sixty-six when I interviewed her, living alone in Florida, and in poor health. So there would be no mistake on Judy's part, her son's wife made it clear that as far as she was concerned, there would never be the closeness for which Judy had hoped for almost half a century. "My husband already has a mother, he doesn't need another one," she told Judy.

The glitter given off by the CHAD Golden Ghetto could be hypnotic, at times creating dreams that rested on shallow foundations. The unreality created by this glitter made it easy for young French women to imagine that their boyfriends were not mere flight line mechanics at Déols, but high-ranking engineers. It had become a truism that even casual glances at life within the Golden Ghetto and the people who lived there provided fertile grounds for a metaphor. Earlier, an admiring Frenchman summed it all up by exhorting that Americans were "rich, rich, rich!" America was the tall blond airman who always seemed to smell better than the young French guy down the street. America was the pile of military script that appeared at downtown bars every pay day, much to the delight of the hostesses and their bosses. America was the wide, shiny smiles that came along with every blue-eyed face. America was the shiny, blue, 1957 Chevy convertible.

The French readily accepted the perception that anything as beautiful as a 1957 Chevy convertible was made to conquer the road, but as Gene Dellinger discovered, not French roads. It was another example of how the Golden Ghetto's glitter, and the perceptions it created among the French easily trumped reality.

"The American cars were big and the roads were small and bumpy, so we had constant problems with everything from headlights to the more serious tie rods, the brakes, and the shocks. Despite how glamorous they looked on the outside, American cars were in horrible shape. Every American who came to Châteauroux was made to understand that he would have at least one accident during his tour of duty.

"To address this problem, we had a maintenance shop in the garage that was run by the base PX. I was selected to do the vehicle inspections for one month, and with two GIs and two French mechanics, it took me only three days to find out that the inspection program was not working. For example, the headlights of some of the American cars were so badly out of alignment they pointed almost straight upward instead of straight ahead. We did the best we could, but American cars were not built for French roads.

"You needed to pass the inspection in order to get your car registered to drive. If we found something wrong, you knocked them down, and that included everyone from enlisted men to full bird colonels and the ladies.

"To show how bad things really were, out of every fifty cars we inspected, I would say no more than 25 percent passed. That's right. You could expect 75 percent of the cars to flunk."

Rank had no privilege when it came to the tough inspection standards set by Dellinger and his four-man inspection crew. Ranking officers and enlisted men would find themselves "without wheels" if their cars were among the 75 percent which failed to get the required windshield registration sticker.

THE FORD VEDETTE, GOOD TEETH, AND ALL THAT JAZZ

Monique Madrolle was ten years old when she was introduced to American-style chewing gum. Even then, she knew value when she saw it. Eventually, Staff Sergeant Harold Snyder and his wife became tenants in the house Monique's parents were offering for rent. The Snyders had a son and a daughter, and Monique found herself earning money as a babysitter without having to leave the family homestead. And, as Monique describes the next two years, it was not just the money earned by babysitting that shaped her perception of just how rich even non-commissioned officers could be.

"Sergeant Snyder and his wife were one of two couples who played poker two or three times a week, and they asked me to babysit the children. Sometimes, they would come back late, and that's when Harold would come home with his pockets full of money. He would reach in and pull out a handful of bills and give them to me. A real friendship developed. They trusted me.

"One time, they wanted to go to Switzerland for a vacation they had been planning for a long time and they took me along to babysit for the children. I was twelve years old. We made many stops in Switzerland. They owned a Ford Vedette, light green, the most beautiful car in Châteauroux.

"While we were on our trip, they decided they wanted to buy me a watch. Even at twelve I had good taste, so when we went to a jewelry store I picked out a very expensive watch. It was a gold watch, real gold.

It cost around twenty thousand francs, that's up about five hundred or six hundred dollars today. They had just enough money to get home. But they kept their word.

"When we got home, I found out that they had to cut their vacation short because of the watch. They told my father, and when he saw the watch he said, 'oh my God, this is very expensive.'" Monique's father reimbursed the Snyders, but too late to save their aborted vacation.

You would not have to look any further than the sequence of events described by Monique to put together a near perfect package depicting French perceptions of Americans. They were generous, almost other-worldly, leather-clad creatures who looked like they "came in from the skies on motorcycles," played poker three times a week, passed out handfuls of money, drove "the most beautiful car in Châteauroux," vacationed in Switzerland, and kept their expensive promises.

But if you were to believe Charles "Josh" Getzoff, a dentist at the base, Americans, with all of their money, had become adept at camouflage. Getzoff, commissioned a captain upon graduation from Temple University Dental School, provided another metaphor that "America is good teeth," but even this could be an illusion.

"I found that a lot of the GIs from rural parts of the South, eighteen to twenty years old or whatever, had already lost some of their teeth. Their dental hygiene was poor," Getzoff said. "We did the best we could, but we were limited to what we could do, crowns and dentures, that kind of thing. We were not permitted to take care of dependents; that was the rule. Supposedly they were to go to French dentists, and God help them.

"One day a year we visited the French dentists. I guess it would be to further Franco-American relations. I went down there and knocked on the door; no one answered.

"The French dentist finally came to the door, and to my amazement, he wasn't wearing a lab coat or anything. He was wearing a butcher's apron and he had blood on it. And I said to myself, what the hell is this? But he did wash his hands. The French used to have an expression when going to the dentist, *je vais á la boucherie.* 'I am going to the butcher.' I understood immediately. That's why we took care of the dependents even though we weren't supposed to."

Janine Artaud was a Châteauroux teenager for whom the acceptance of an illusion was made easy. She had met Hollywood-handsome Jim Hawkins, fallen in love, married, had a son and a daughter, and embarked on a fairytale life that took shape in 1956 when she and Jim attended a CHAD concert by the silky smooth Platters. Their blockbuster hit, "Only You," was to shape their lives, and preserve a lifelong love affair.

Janine was lucky, because in her case, her husband was up front from the very beginning when they were already deeply in love. "Jim told me he was poor. He never lied to me. But I didn't care because I loved him so much." Jim's acknowledgement that he was poor was supported when the couple arrived at Jim's home in backwoods Kentucky, a tobacco sharecropper's cabin in which she was made to feel unwelcome from the very start. The welcoming meal was a bologna sandwich. That bologna sandwich symbolized what the future would hold for them if they were to remain in Kentucky. He returned with his young family to Châteauroux, where Janine's parents accepted him. It was already apparent that Jim's health was deteriorating, his breathing was becoming labored, but this signal was largely ignored.

Jim and Janine returned with their young family to Châteauroux. "Jim got a job as a bartender at the base. He felt that being a bartender at the base was a dead-end job, so he decided to go back into the American military," said Janine, who was able to get a job as a translator at the Déols air station.

"Vietnam was raging at the time. Jim's health was not that good. He was unable to get back into the Air Force, but the Army accepted him. We all went back to the States with him after he enlisted," said Janine.

Jim Hawkins had served his country honorably in the Air Force, returned to civilian life, and when he attempted to reenlist, the Air Force turned its back on him. But his bad lungs and labored breathing made little difference to the U.S. Army, which needed a man who could carry an M-16 and wade his way through the stinking jungles of the Mekong Delta.

Janine followed Jim to his first posting in Germany just before he was shipped to Vietnam. Janine and the kids remained in Germany. Châteauroux was a magnet for her happiest memories. Those early days when as a teenager she and Jim were captivated by the Platters', "When you hold my hand I understand the magic that you do. You're my dream come true, my one and only you." The young couple embraced the American ethos with jazz as the driving force. "We loved American jazz. One time, Jim and I went to Vierzon, which was not far from Châteauroux. We went to hear the great American jazz artist Sydney Bichet. It was a wonderful time. He was a great saxophone player. You got the feeling that you were right in the middle of American jazz. The room was so dark and so smoky; it was wonderful. You felt very free, there was a feeling of a lot of freedom.

"Only You" became our song. It was always our song even after our divorce, every time I wrote to Jim, I would sign off the letter with the words, "Only You."

Janine, Judy and Jeanine exemplified what I found to be a minority among the dependent women I interviewed. Their free and unadulterated use of the word "love" was at the core of their stories. There was never a sense that intoning their "love" during their interviews was perfunctory or obligatory. But with others, it seemed more often than not that it was easier for young French women to speak of their "boyfriends" as the door openers to a "new life," a means to an end.

Stories like these were hardly a rarity in Châteauroux during its heyday. There were surely opportunities readily available to young, wide-eyed Americans, many of them away from home for the first time, who were invited into a world they had never known existed. It came down to a matter of booze, women and erotic expectations.

1959 – COLD WAR POTPOURRI – 1959

- Military advisors Major Dale R. Ruis and Master Sergeant Chester M. Ovnand became the first Americans to be killed in Vietnam when ambushed by communist guerillas in Bien Hoa, just outside Saigon.

- French directors such as Jean-Luc Godard, François Truffant and Alain Resnais introduced New Wave cinema with films like *Hiroshima Mon Amor* and *Breathless*.

- February 3rd was "The Day the Music Died" when rock stars Buddy Holly, Richie Valens and "Big Bopper" Richardson were killed in a chartered plane crashed shortly after takeoff from Clear Lake, Iowa.

- On March 9th, the public met the Barbie doll for the first time, buying 350,000 Barbies without knowing that the Mattel Toy Company used as its inspiration a German cartoon prostitute called Lilli.

- A pair of Cold War friendships was initiated in the Middle East when the United States signed an economic treaty with Iran and the Soviet Union signed an economic and technical treaty with Iraq.

- Women could breathe easier as they discarded their girdles and garter belts thanks to the introduction of panty hose by the Glen Raven Mills in North Carolina.

13.

14.

15.

16.

17.

18.

19.

20.

21.

22.

23.

24.

NOTES TO PHOTOGRAPHS 13–24

13. First Marriage, 1952. Airman identified only as Sgt. Martin receives his marriage certificate from a city official. His French bride is not identified.

14. Stripper Entertainment. This stripper was one of several that came in every week from Paris to perform at the CHAD Officers Club and NCO Club, complete with a four-piece band and vocalist. These were stag performances, note the all male audience. Photo courtesy of Yves Bardet.

15. Black and White Partygoers. A scene hardly ever seen in small town mid-America during the 1950s, black GIs and French country girls party in a suburb of Châteauroux. Photo courtesy of Janine Meriot Anderson.

16. Staff Sgt. Lawrence Anderson and his French bride Janine Meriot in her wedding dress. Janine, traveling across America from one Air Force base to another with Lawrence and their two children, experienced what state side segregation meant. Their marriage endured more than 50 years. Photo courtesy of Janine Meriot Anderson.

17. Jim Hawkins and his French wife Janine Artaud. Hawkins, described by Janine as "Hollywood Handsome," was the son of a Kentucky sharecropper. Exemplified that not all Americans were rich, rich, rich. Janine was welcomed into the Hawkins family by Jim's unsmiling mother who offered the couple baloney sandwiches for their first meal. Photo courtesy of Janine Hawkins.

18. Joe's from Maine restaurant, a virtual home-away-from-home for lonely GIs serving everything from hamburgers, BLTs to french fries, beer and ham and cheese sandwiches. Also something of a marriage mill where possibly as many as 20 of its barmaids married GIs. Photo courtesy of Mike Gagné.

19. Sam Herrera and Anna Reh (center), courting days in Châteauroux 1958. Herrera escaped from a Colorado coal mine to join the Air Force. Anna was carried on the back of an older sister from communist Yugoslavia. They met at CHAD where Anna worked as a nanny for Sam's boss. Photo courtesy of Sam Herrera.

20. Base Commander General Pearl Robey and family in residence. Communists were quick to point out the American officers commandeered many of the Châteauroux area mansions. This photo taken in the reception hall of their mansion in St. Maur did little to dispel these claims.

21. Animal House. The hunting lodge of Ferdinand de Lesseps, the builder of the Suez Canal, was transformed for over a decade into the favorite den for partying

and trysting by a select group of high level CHAD officers and civilian employees. Author's photo.

22. General Red Forman Chateau. The former mansion of a chateau industrialist, this chateau shared almost equal billing with the Ferdinand de Lesseps Hunting Lodge for fun and games. Forman's three children shared in the hi-jinx by surprising guests with ample use of a reverse flushing toilet. Author's photo.

23. Red's Fun and Games Clique. After a day on the Swiss ski slopes, General Red Forman (center), relaxes with his entourage. Donna Hildebrand, far left, and his wife Boopy Forman, far right, bookend the group. Their partying was mythic attracting high ranking officers such as Lt. General Glen Birchard, second from left, to join the fun. Photo courtesy of Boopy Forman.

24. Officers Club Hi-Jinx. JAG Lt. Bob Goggin was the star of this toga party, Donna Hildebrand on far left, Kay Wood second from right all of them stalwart members of Gen. Red Forman's fun loving clique. Photo courtesy of Boopy Forman.

CHAPTER 9

CHE GUEVARA AND A SEWING MACHINE

The Homelike Atmosphere to Suit Your Mood
Bar * Lounge * Dancing
Striptease Friday–Saturday–Sunday
—Jimmy's Club appeal for new GI members

The clarion call trumpeting the virtues of Châteauroux's Golden Ghetto was loud and clear enough to be heard across the Atlantic Ocean. Once the pilgrims had arrived, they were greeted with erotic enticements such as those offered above by Jimmy's Club. Every mood would be catered to, and strippers would make sure members were not bored on weekends. Located at 12 Rue Paul-Louis-Courier, Jimmy's was perfectly situated within a compact maze of downtown Château-roux streets where it was easy for young GIs to walk or stumble from one bar to the next. Luring these young guys through the front door was not enough—you had to make them feel like they belonged, had a piece of the action. Membership clubs became the rage. A young American could now squeeze himself onto a cramped barstool next to his favorite hostess, secure that the few bucks he paid for his membership card was well worth it.

These cards provided proof that fractured English syntax, as well as mystery words, made little difference as long as the bucks kept coming. The Crossroads Club, 38 Avenue de l'Hôpital, and Goldy's Club, 94 Avenue de l'Hôpital, shared the same rules.

Membership available to all by approval of applicants by the Board of Directors. Cannot be coming to a lady and member will result in membership being termi-nated. No fighting at any time. Intoxicated persons will not be served. It is the owner's privilege to pull any members card, strike at card, due to bad conduct.

For as little as a buck, these crudely produced membership cards were able to secure a rare sense of belonging for young airmen thousands of miles from home. Sure, Madame Janine and her cohorts were in it for the money, but they also helped provide these guys with some of the more memorable moments of their lives.

Patrols by the Air Police could be expected every fifteen minutes, more or less, and this provided a window of opportunity for any drunken brawls that might have been festering. There were plenty of hostesses, prostitutes, and working ladies masquerading as "students," to make bar owners circumspect. It did not take much imagination to figure out why Jimmy's, the Crossroads Club, and Goldy's Club warned members that "personal and business affairs will be subject to approval by the Board of Directors."

Châteauroux's bar scene was a parallel universe. All were welcome at Madame Janine, Second Blvd, Saint-Denis, *Le Chat Qui Fume* [The Smoking Pussy] at 7 Rue Gutenberg, the Frog Pond, with its two up-stairs rooms available to working ladies, and the Beer Row line-up of bars along Rue Ste. Luc. Black airmen had their favorites—the Jockey Club and the Crazy Bar, one of them in Ste. Christophe just outside the Châteauroux boundary. The legendary Joe's From Maine on Rue de la Poste was still going strong, proving that French fries, chili, and hamburgers could provide a balm for even the stormiest of Franco-American wounds.

Pilgrims returning for the 2007 reunion clearly identified their cultural preferences. Never mind that if they were to stop at No. 4 Rue Descente de Ville they would find a sixteenth century toll house that still beckoned with myriad windows, buttresses, and a turret with a pointed roof. Instead, they were shocked and somewhat disbelieving to find that the Crossroads Club was now a small parking lot. There seemed to be little interest in the early thirteenth century St. Andre's church on Rue Alain-Fournier, or in its long history. Originally the Church of the Cordeliers, it had survived the Revolution, a secular takeover that renamed it the "Temple of Reason," and finally was returned to Catholicism in 1876. But there was dismay that the Frog Pond was now a small apartment building with an out-of-business video store on the first floor. On the corner of Rue des Notaires is a sixteenth century bourgeois dwelling called the House of the Cadran, but this could hardly compare with the fact that Jimmy's Club is now an ordinary dwelling and that Le Lido now offers fresh fish, beef, horsemeat, and an endless list of edible goodies as Les Halles in downtown Châteauroux. A short survey conducted during the 2007 reunion failed to uncover anyone who knew that the third of nine Song of Roland manuscripts in Old French is housed in Châteauroux's Municipal Library, as are manuscripts handwritten by Napoleon Bonaparte while on Ste. Helene island that describe his campaigns in Syria and Egypt. But there was great interest among a group of old airmen as they laughingly compared their saloon membership cards, and their half-century old memories. They could have been standing on any street corner in the United States comparing baseball trading cards; just kids who had returned to Shangri-La for three days of youthful exuberance.

Implicit in many of my interviews with Americans was the shared opinion that Châteauroux was a gray backwater town that offered little more than bad weather and almost half a population that did not want them there at all. It was viewed as an ugly duckling that had managed to hide its impressive pedigree. There was little evidence during the CHAD years that Châteauroux was, for many decades, a prize that was fought for over and over again. In 1356, Prince Edward of England, the feared Black Prince, failed to capture the fortress chateau, and exacted his vengeance by burning the town to the ground. By 1441, after decades of being ravaged and pillaged, Châteauroux became a city protected by high walls and fortified towers. The town was still coveted centuries later when General Hicks laid claim for the Americans. A town that would not fall to England's Black Prince almost six centuries earlier despite all the bloodshed willingly gave itself to the Americans without shedding a drop of blood.

Many Americans expressed regret at having never taken the time to discover that the Golden Ghetto had indeed extended its polished tendrils into a countryside that was well worth exploring. Josh Getzoff, who earlier expressed his astonishment upon encountering a Châteauroux dentist at his office clad in a bloody butcher's apron, had vowed he would have no regrets. Châteauroux's clarion call beckoned him with increased intensity during his final two years at Temple University Dental School in Philadelphia. When he arrived at the 7373rd U.S. Air Force Hospital in CHAD in 1960, Getzoff was well primed, with his expectations running high. Promoted to captain shortly after his arrival, he began to scope the terrain in and around Châteauroux and the air station to see if what he had been told could possibly be true.

"I chose the Air Force, and I chose Châteauroux because there was a fellow named Ronnie Gross. He was two years ahead of me in dental school and lived in the same neighborhood in Philly," said Getzoff. "He was in Châteauroux from 1958 to 1960 while I was slaving away in dental school, and he was writing all these wonderful letters about wine, women, song, and travel.

"After dental school, I knew I had to go into the military. With Gross' words ringing in my ears I put in for the Air Force. They asked you to put your preference for stateside duty and overseas. I was given three preferences, and I just put down on each one of them 'Châteauroux, Châteauroux, Châteauroux,' and I got it!"

It took only two weeks for Getzoff to realize he was not going to live at the base, deciding instead to live on the economy in town. To do this, he needed a car, so he bought a Peugeot 404. Then, he was lucky enough to find a vacancy in an apartment that he shared with two other officers on Place de Impasse Charlier close to the downtown area.

"Châteauroux was a real eye opener for me. When I got there I was twenty-three years old, and it was amazing to me what these young women, eighteen and nineteen years old, knew about sex. Remember, I was a child of the 1930s and was a teenager in the 1950s, which was basically an innocent time.

"After a couple of weeks I was feeling my oats. I got a date with this young lady and we went up into the bedroom, and I made my first mistake, which was that I tried to put a condom on. This was a habit because back in the States the pill hadn't come around yet, and you didn't want a pregnancy so this is what you did. This young lady looked at me and saw what I was doing and—I'm not kidding now—she looked me in the eye and said, 'if you keep wearing those, you'll never get a date.'

"Let me tell you this, that after this first experience, I always had condoms with me but I never took them out, and never had to use them during the three years that I was there. It was also an experience, watching her use the bidet for the first time. I didn't know what the hell a bidet was. I just came from Philly and what the hell did I know about a bidet?"

Thomas Young, who marveled at how seductive young French women could become simply by sporting a scarf, a ribbon, and a home-sewn pleated skirt, had succumbed to the erotic appeal of Châteauroux in 1958, and found it almost impossible to stay away. He did not share Getzoff's sense of incredulity upon discovering how much expertise could be found in such young packages. "The French have 'been at it' for close to a thousand years—that's five times longer than the U.S.A.—I would submit that there is a lot of catching up to be done, 'back at the ranch.'" A part-time expatriate with a home in Amboise, he remembered his first impression when he arrived at CHAD as a civilian employee.

"Even blindfolded, you can always tell when you're in Châteauroux, just sniff the air and listen to the sound of purring. Châteauroux and its surrounding villages and countryside became a super-heated cat house because of the combining of several important factors: NATO forces (80 percent American) brought tons of money and vanquished local boredom; and after six years of war (and war restrictions), plus a struggling recovery and rebuilding effort, the young French girls were hot to trot, day and night."

By 1960, Châteauroux had developed a reputation that transcended the U.S. military's class system; the word was out, and passed along between enlisted men and officers alike.

A Little TDY Ditty

It made little difference that Dover Air Base in Delaware was thousands of miles from Châteauroux. Jay Parsons explained that there was a linkage between the two bases that was forged and strengthened by expectations that were always realized, often to the point of wonder by those involved. Parsons, an Air Force career pilot, served at CHAD from fall 1961 until 1966 when the base was about to close. Before becoming an aide to Brigadier General Robert D. Forman, Parsons piloted C-118 surveillance aircraft that were in the air over Europe twenty-four hours a day. But it was his experiences as the pilot of the huge C-124 cargo planes that led to fond memories.

"The crews would be on temporary duty while at Châteauroux after they arrived with their load of cargo. We would then send them to other bases across Europe, to the Congo, maybe to North Africa, to Beirut, Lebanon. Wherever we had to ship cargo, that's where they went." The cargo planes always returned to Châteauroux after their supply missions, and this is where the crews wanted to be as the air base reputation for fun and games grew. "Châteauroux really had a reputation among the pilots. You always knew you would have a good time there, plenty of women. In fact, there was even a little song about it, how great it was to be assigned TDY [temporary duty], generally for ninety days after you flew in. It was just a little ditty and we sang it just as you would sing that kid song, 'Bye-Bye Blackbird.'"

After laughing his way through two attempts, Parsons took a deep breath and sang:

> When there's nothing else to do, it's ninety days in Châteauroux.
> We fly fat birds. All day long we sit around,
> We don't fly until the sun goes down.
> When the days at Châteauroux are over,
> We fly fat birds back to Dover.
> I'll see you and you will see me,
> We'll all meet at the TTC. It's fat birds we fly.

The "fat bird" was the huge C-124 that made regular runs between the Military Air Transport Service (MATS) base in Dover, Delaware, and CHAD. The TTC was part of the main base area at Dover that include the officers' club.

Parsons was a sophomore at Texas A&M University when he and twelve others dropped out of school to join the Air Force Cadet program. He was only thirty-nine when he was forced to retire after more than twenty years as an Air Force pilot. The most memorable of those years were when he served as an aide to General Forman with the 1602nd Air Transport Wing. He never rose above the rank of captain. During his time at CHAD, he became a member of the clique of competent, hard-working officers and ranking civilian employees who surrounded General Forman and his wife, Boopy. Its members somehow managed to get their jobs done despite its legendary hard drinking, hard partying, and hard playing. As it was with Getzoff, living at La Martinerie after becoming Forman's aide was just too restrictive for the fun-loving Parsons, who would eventually marry four times. "Châteauroux really had a reputation among the pilots. You always knew you would have a good time there, plenty of women," Parsons said. Before his permanent assignment at CHAD, he was among those TDY pilots who shared a preferred status among the young ladies.

Marty Whalen, who had been pulled from his pilot seat and ordered to take over the financially troubled Officers' Club, said, "Well, first of all, single women didn't hold the single officers at the base in very high esteem. The women thought they were very cheap. The officers who were bachelors very seldom invited young single women to dinner. Instead, they would say, 'let's get together tonight and I'll meet you at the club at eight-thirty.' Or they would just say, 'see you at the club.' The guys who came in TDY were a different story. They would fly in, pick up some girls, have drinks and dinner and then all kinds of stuff. I'm sure that not all of them were bachelors, probably a lot were married with families at their home base. But when they came in, they still treated the single women very well."

Châteauroux was like catnip for Parsons. There were plenty of women, and one in particular brought back fond memories. "I shared an apartment with another officer in downtown Châteauroux. I guess you could call it a penthouse apartment. I don't remember the street address, but it was directly across from the tallest building downtown. We even had a small grass lawn that had to be cut.

"I had a young French girlfriend, really beautiful, who would visit me. As a pilot, I was out of town a lot, and if I wasn't there, she would leave a note on the door, signing it 'À Bientôt,' which as you know means 'I'll be seeing you soon.' When I got back into town, other residents would smile when I asked them if I had any visitors when I was out of town. They would say, "'À Bientôt was here,' sort of like 'Kilroy was here,' which was left all over Europe during WWII."

For commissioned officers, living on the economy and having your own apartment in Châteauroux had definite advantages. As Getzoff discovered, it was not only the bachelor officers who valued them as hideaways. "I was there maybe a month or two when the first C.O. I had, Colonel James Long, came up to me and said, 'I want to see you in my office.' And I said to myself, 'what the hell have I done?'

"First of all, Colonel Long was from Tennessee, and if you were black, forget it. He called me 'boy.' I'm not black, but he called me boy, which irritated the hell out of me. I went in there, and he said to me, 'Are we going to have the same arrangement as the other guy?' I didn't know what he was talking about. The arrangement was [that] I gave him the keys to my apartment, and on Fridays he would use my apartment. He was with one of the other colonel's wives. Is that a bad thing to say?

"I could have written the script for the movie, The Apartment, that was released the year I arrived at CHAD, with me playing the Jack Lemmon part, and Colonel Long playing the Fred MacMurray role. That slob never cleaned up. All my sheets would be messy.

"It was a sensitive matter because I knew the other colonel whose wife Long was bedding down in my apartment. I played cards with her husband. I always wanted to tell him, but I didn't. I do not want to mention any names."

Even among the Americans who admittedly were always on the prowl for females, there was always the blissful chance that they would run into a French maiden who was extra special. For Parsons it was A Bientôt, and for Thomas Young, it was Gi-Gi, a teenage bartender who captured his imagination from the very beginning. He described one blissful night spent sharing her small room. "After what the French call 'an unforgettable moment for all,' we slept. I awakened to minus ten degrees Celsius in her small third-floor room, which had no heat (except ours). I wondered how we would survive once out from under the quilt."

Young described how Gi-Gi sprung from the blankets and went into action. "Gi-Gi said in her best English: 'No sweat, G.I.—you watch.' My eyes were glued to her lithe, nineteen-year-old's body. Grabbing a round, enamel basin and a corked bottle, she poured liquid into the basin, stepped back a respectful distance, and tossed a lighted match its way. With a tremendous 'whooff,' blue flames grew out of the basin, soon curling the paint on the low ceiling, and the room became as warm as toast in five minutes. 'You see, GI,' said Gi-Gi, 'French technique!'"

Not even political ideology could interrupt the flow of progress across Châteauroux's sexual landscape. Getzoff was well aware of the Communist Party's strength in central France, and found that you could discover a Communist in the most unlikely of places. "One of the French ladies I was dating, spending some real quality time with at her apartment, was a Communist. So if you can imagine this, here we both were, stark naked, really going at it, and looking down at me were two posters of Che Guevara and Vladimir Lenin. It was very interesting."

Three-man basketball games in which he was the only white player expanded Dave Bankert's athletic horizons, and it was the discovery of a little known French law that excited his teenage erotic fantasies. "The law said that you had to be eighteen in order to get an international driver's license. Well, we uncovered a French law that said if you had a motor scooter with under 125 ccs of engine size, you didn't need a license. During the summer, there was only a handful of us around at the base. The rest, Air Force, Army, and civilian kids, returned to the bases their families were stationed at. So we all went out and got Vespas; we were called the Vesparados.

"The enlisted men had to be back in their quarters on the base by midnight. There was a curfew, at least at that time. This did not apply to us kids. So at eleven-thirty we would show up on our Vespas at the various bars and dance halls downtown. We had these motor scooters so we could really get around town evaluating the action at the various places.

"It was easy pickings, there really wasn't very much for us to do but show up. There were hardly ever any men left in these places, but plenty of women. They weren't kids, probably twenty-one or twenty-two, and some of them older.

"Many of the women had been drinking all night, and were feeling quite friendly when we got there. We would go in and order drinks. It didn't make any difference how old you were, you got served. I remember I always ordered a gin fizz without really knowing what it was. But I did know that meant exactly the same thing in French and English. Why complicate things?

"So there we were, a bunch of teenage kids with older women readily available. It seemed that in Châteauroux the men were either too old or too young, and this opened up some great possibilities for us. There were a lot of young women, some of them really beautiful enough to be movie stars."

If Bankert's late night and early morning exploits were testosterone-laden, there was a French high school kid who, fifty years later, still viewed the downtown saloon scene as unforgettable. Leandre Boizeau might have distributed anti-American pamphlets for the local Communist party, but this did not preclude him from taking a youthful peak at the action. "When I was in school, every Saturday we would see the trains filled with prostitutes," Leandre Boizeau said. "We'd go see them at the train station. There were all the American bars, Jimmy's, etc. It created a considerable nightlife. There were the MPs who kept track of the GIs, knocking them on the head if they were out of order. It was unbelievable. Once we saw them, we knew that was it for the night for those GIs."

AMERICAN BARS, BLACK AND WHITE

Irene Smith, the enigmatic French woman who had successfully hidden her marriage to a black GI for more than forty years while retaining his non-French name, was convinced the Americans had brought racism to Châteauroux. If so, then the almost ritualized self-segregation that prevailed both at CHAD and in town was a natural appendage, one that was accepted and even lauded if racial peace was to be maintained. Gene Dellinger was among the first arrivals in 1951. "In the beginning in Châteauroux, there were three bars that had different GI customers. The Jockey Club was for the blacks; the Le Lido was for both blacks and whites, a dance hall that was mixed; and the Tivoli was for the whites. The whites did not go into the black bar, and the blacks did not go into the white bar. It was an unwritten rule that meant 'stay away.'

"If you walked into the Jockey Club on a weekend, you could look around and then you were expected to leave, no problem. It was the same thing when blacks walked into the Tivoli. They walked around and wouldn't stay long. This was in 1951–53. When I came back for my second tour in 1956, the same situation was still there."

Describing prostitutes in Châteauroux during the 1950s and 1960s depended, to a large extent, on the vantage point. Communists, most notably in their newspaper, *La Marseillaise*, were fond of branding as prostitutes any French woman who frequented downtown "American bars." As we have seen, many of these young French women had come to town for a good time.

To describe this enticing arena, it would be hard to find anyone with more hands-on knowledge of Châteauroux's downtown bar, restaurant, and prostitute scene than Fernand Marien, also known as "Gypsy" or "Bambi." He had owned several hostess bars, and from these vantage points had witnessed the ebb and flow of nighttime action. Excitement was at its peak during the middle and end of each month—paydays for American airmen. The bi-monthly migration changed very little from one payday to the next. "The buses that came from the base in Déols or La Martinerie all stopped at the railway station," Marien said.

"So those who had no cars had to walk down Avenue de la Gare. They went from [hotel] Le Faisan to Place de la Marie. Later, when many of them had cars, they scattered throughout town. Women came from Orléans, Tours, and Paris. This was for the hostess bars. Actually, they were not really prostitutes, they were hostesses. The real prosti-tutes came from Paris. They came every two weeks, the day when men were paid. They would stay here for a couple of days. And they walked the street, mainly in Avenue de la Gare. I remember we counted up to eighty-four in that street, Avenue de la Gare. Americans mostly went to Avenue de la Gare, or Place Voltaire or Avenue de Verdun which is also called Avenue de l'Hôpital, where there used to be many bars. We could see the prostitutes in the street where there were many hostess bars. I remember counting up to twenty-four hostess bars."

Gerry Lowry joined the Air Force immediately upon graduation from high school in Rozelle, Maryland. Like many other GIs I approached for interviews, Lowry viewed a four-year enlistment in the Air Force as a neat alternative to getting a notice from his local draft board. "The draft was going on at that time, and I didn't want to be drafted after I got out of high school in 1955," he said. At CHAD, he met and fell in love with Nicole Tuillemain and extended his enlistment in order to marry her before shipping home. From the start, Lowry wanted to fit in, not only at the base, but also with the citizens of Châteauroux.

"Well, I was going to be there for three years and I wanted to make the best of it. I wanted to learn the language, so I went to school for six weeks, and I thought I knew a little bit of the language so I went downtown to try it out. I found out very quickly that the people did not understand me. So I said, 'to heck with this,' and was going to stay downtown and begin associating with the French people who lived there."

Before meeting Nicole, Lowry was no stranger to Châteauroux's downtown bar scene, with all of its erotic enticements. Jimmy's Club was his favorite, but he also made the circuit to other GI saloon mainstays like the Crossroads and the Frog Pond. "I never went to any of the black bars because they went to bars by themselves. I don't remember the names of any of these bars, because as I say, I never had any desire to go to any of them," he said.

"I would go downtown on payday. We would sit around at an outside café when the prostitutes came around. I only knew a little bit of French, but what little I did know I was able to have a little bit of fun with the prostitutes. When one of them would approach, I would say 'Combien vous me payez pour rester avec vous?' (How much would you pay me to stay with you?). She would laugh and walk away and say, 'Vous avez été en France trop longtemps!' (You have been in France too long!). She was like almost all the other prostitutes, pretty. They were what you would expect as registered prostitutes. They all had green cards. You'd hardly expect them to be ugly."

Patrolling the downtown bistros, hostess bars and otherwise, was a major part of Staff Sergeant Willie Ward's duties as an air policeman. "We had ours and they had theirs, this was by choice," said Ward, whose favorite was the Jockey Club. His description of the club could easily leave a listener wondering if the white airmen realized what they were missing. "We went there fairly regularly. It was a typical bar with booths and tables on the right, and a long bar on the left, and in the back there was a room for dancing with live bands. There were some great musicians that could play anything.

"The musicians were black, French, and one guy who was a member of the band squadron at the airbase who would sit in every once in a while. They'd play jazz, blues, popular dance music, and rock 'n' roll."

There was one legendary prostitute whose name sifted through more than a few interviews. Nobody knew her given name, except for the one bestowed on her by the GIs, "Nine Fingers."

"It's kind of hard to describe her. Remember, this is more than fifty years ago. I believe she was from the area," Sam Herrera said. "As I remember, she was older than the other locals, and she was clever. She had only nine fingers, I believe the index finger on one of her hands was missing.

"She was very clever, especially when it came to the younger GIs, the guys who were really naïve. From what I recall, they would get together and agree on a price. She would always be paid in advance. Then she'd take them to a place to do their thing, he'd take off his pants and underwear, and then she'd run away. They could hardly run after her, could they? I know that she worked all over town. I knew one of the guys, I guess you could call him one of her targets, very well. He was a guy that I worked with. We all called him 'Myrtle,' so you can guess how sophisticated he was."

It was interesting that in 2006, while interviewing Joe Gagné, the owner of Joe's from Maine, a restaurant and bar located almost squarely in the middle of the downtown fun zone, he voiced no moral indignation over prostitution, but became irate about how bad it was for business. "For a long time there was prostitution in the downtown area. They used to come in by train from Paris. They would come into my place, and I could see they were taking my customers away. A prostitute would come in and have one beer with a GI and then leave with him. Usually, a GI would have four or five beers. It wasn't good business. This had to be stopped so I put a sign outside that said, 'Unescorted women not welcome.' That kept most of them away. The ones that did come in, I threw out. After that, it was GIs coming in with their girlfriends."

TICKETS TO *LE BAL*

It was not a surprise to learn that things did not always run smoothly in a heated environment of women and booze. Club membership cards warned that drunken mayhem would not be tolerated, and that if found guilty by a somber, dually appointed club membership board, American airmen faced the possibility of losing their membership status and the one or two bucks they paid for it. And it was not always those hard-drinking, hostess-hunting GIs who caused the problem.

Joe's son, Mike Gagné, recalled that competition among downtown bar and restaurant owners was sometimes intense, and could lead to an improbable confrontation. One involved Gypsy Marien, his one-legged brother, Clovis—better known among the downtown regulars as "*patte de bois*" (wooden leg)—and Joe's dog. "Clovis ambushed my father and one of his employees when they were on their way to the Frog Pond to recover my dad's dog. Then Gypsy and Clovis pulled Joe's coat up over his head and beat him up. But Joe got the dog back.

"This did not mean they were enemies or anything like that. They've seen each other socially over the years, and it seems to me that they are on friendly terms."

Thomas Young described how even the air police, protectors of the innocent and enforcers of the law, could be unwittingly drawn into jealousy-driven and potentially dangerous outbursts. In this instance it was in *Le Chat Qui Fume*, a very popular bar that was regularly checked by a two-man patrol that included an air policeman named Paul H. Paul. He was "a handsome devil from South Carolina, who was married to a Southern Belle with class, and she became more and more upset over the degree of success Paul was having with the French girls in the night-spots he patrolled. All their eyes were on him when he walked in."

"Their off-base housing being one hundred yards from *Le Chat Qui Fume*, she went over the top one evening, somehow obtaining Paul's Colt .45 automatic, walked into the place and shot up the ceiling. Plaster, dust, and general panic everywhere. Everyone dived under tables, Blondy (the owner) called the city police who, informed of the ongoing shooting, said 'call back when it's over—we'll come and make a report.' Paul's wife walked calmly out of the place and back to their home. Paul and his lovely spouse were 'shipped back early,' back to South Carolina."

Borrowing from Jimmy Reed's blues classic, "Bright Lights, Big City," I found it hard to imagine that there was not at least one kindly, big-hearted madame among the mix of characters who brightened Châteauroux's once drab downtown streets with incandescent glare. Enter Madame Lydia.

Alain Birckel was only fifteen years old when he found he had a natural instinct for troubleshooting and repairing British sports cars. The Golden Ghetto offered a hard to imagine opportunity to the son of a Communist railroad worker, who had been raised in a one room Déols home with no running water. After seeing what little miracles Birckel had performed on his British Triumph, an Air Force transport pilot named Major Cox agreed to sponsor Birckel to permanently move to the States. This was in 1966 after Birckel had fulfilled his French military service requirement and returned to Châteauroux. Cox had spent a year and a half taking care of all of the paperwork, Birckel just had to find a way to get to Paris where an airplane ticket awaited him. Madame Lydia came to his rescue. "I met her when I was about fifteen years old and working in a garage at that time. The prostitutes worked at various bars around town because they needed a place to operate, and at these bars there was always a guy whom I guess you could call a pimp. There was this woman, Lydia, who had places in Paris and in Châteauroux and she would be here on weekends. For some reason, she liked me. She's a very nice lady. She was very warm, very professional.

"If you were to ask me, I would say that she ran prostitutes out of her business in Châteauroux. I think most of them came from the Châteauroux area, and they lived here. But she liked me and she was good to me. She arranged for me to be driven to Paris and for a place for me to stay until my flight to the States.

"You couldn't really say we were friends, but she did like me and she knew this was going to be my big chance to better myself," Birckel said. "I never asked for her help. She just offered it. Her help changed my life." Birckel spent more than thirty years in the United States rising to service manager at one auto dealership before becoming sales manager at a Toyota agency in Denver. Cash flow problems brought on by two divorces landed him back in Berry where he admitted being "like a fish out of water."

Châteauroux's downtown prosperity was due to a sometimes stormy mix of respectable restaurants, cafes, bars, and shops sharing space with hostess bars, professional street walkers, and "working ladies" insisting they were students. It was inevitable that there would be a denial that this was the real Châteauroux. This was certainly the viewpoint of Gérard Claude Audas. He was born in Châteauroux in 1934, the son of a career gendarme and a teacher. After a posting in Damascus, Syria, the family returned to Châteauroux at the outbreak of WWII in 1940. With professional parents, Audas was bourgeois, with all of the instincts that came with his family's class. After getting his baccalaureate in Blois, he went on to earn a civil engineering degree in Strasbourg, and was later recruited by famous American golf course designer and builder Robert Trent Jones, his employer for ten years. He was a teenager who witnessed the downtown transformation after the Americans arrived. Emotionally, he might not have liked what he saw, but being a practical kid, he was able to rationalize a unique way to put Uncle Sam's money to good use.

"After a while, when you could find perhaps fifteen or twenty prostitutes on one street, it was not a good thing. The GIs got paid every two weeks, and every two weeks the prostitutes would come in by train from Paris. It was an easy thing for a GI to drive by in his car and take them to a bar or hotel outside of town. They did this because Châteauroux was essentially still a bourgeois town," he replied.

"Downtown Châteauroux was a lively place, with a lot of GIs with a lot of money. This was tempting. There were three of us who got our baccalaureates in Blois, and we decided to have what we called *Bal du Baccalaureate* at Le Lido in downtown Châteauroux. But we had to find a way to pay the club, and we figured an easy answer would be the Americans. We approached the Americans because we knew they would want to come to the celebration because there would be girls. The American soldiers were rich to us, so naturally we would ask them if they were interested.

"If you ask me whether we charged the Americans more, my answer would be 'yes, yes, sure, sure.' I would say three times as much. But it was not always the same price for every American. In French we say '*a la tête du client*,' which in France means 'the impression you get from looking at a person,' if he has money or not, or if he has a lot of money. And this was important because we had to pay the money for our celebration. And it's nice when you have some left over that you can put in your pockets."

There must have been something in the thick smoke from American cigarettes wafting out of Le Lido that caused young Frenchmen to rationalize their behavior. James Besset, two years older than Audas, was also born in Châteauroux. Earlier he had recalled the fierce turf battles fought between the Left, and Right Wing *Maquis* during the final months of WWII and immediately thereafter.

"It was a boom time for bars and prostitution in Châteauroux," he recalled. They were generous with prostitutes. I remember Rue Joseph Bellier. I think it was Le Lido. There was a little room at the entrance where you could eat, and there was a bigger room downstairs where you could dance. There was a sort of circular gallery all around with tables, we would dance there.

"One day, with a couple of friends, we had eaten there. There were some Americans at the counter. They were completely drunk. One of them had money in his back pocket and we noticed that the money had fallen on the floor. One of my friends picked up the banknotes and said laughingly that since it was French money, it was our money. So we spent the money drinking what the other guys would have drunk. They were in such a state they didn't notice anything, and we didn't feel guilty because we guessed that they had other banknotes."

Pierre Pirot understood that what Châteauroux was experiencing was natural in a military-dominated town. "There were more and more night clubs where you could find prostitutes. There were fights at night. It was not always very funny, but that's part of their attitude. We can't say that was true for all of the Americans. I remember a GI who really had strong feelings for his prostitute girlfriend. To show his affection, he gave her a sewing machine for which she had little use. As a result, I bought the sewing machine from the American boyfriend. It seems the prostitute could do many things very well, but sewing wasn't one of them."

1960 – COLD WAR POTPOURRI – 1960

- On September 26th, 70 million viewers watched the first televised presidential debate when Vice President Richard Nixon and Senator John Kennedy squared off at a Chicago studio.

- On April 4th, RCA Victor Records announced that for the first time they would release all pop singles in mono and stereo simultaneously. Their first release was Elvis Presley's "Stuck on You."

- Soviet Premier Nikita Khrushchev disrupted a fall meeting of the U.N. General Assembly with shouts and then banging his shoe to show his anger about U.N. intervention in the Belgian Congo.

- On April 30th, eighteen prominent South Vietnamese signed a petition criticizing the U.S.-supported regime of President Ngo Dinh Diem for its corruption.

- In September, the American Football League debuted with eight teams, offering competition to the entrenched National Football League while enticing 50 percent of the NFL's draft choices to sign AFL contracts.

- In November, John F. Kennedy became the first Catholic to be elected president, defeating Vice President Richard Nixon in a closely contested race.

CHAPTER 10

NOMADS AND AIR FORCE ONE

Just about all the students were former GIs. I was there in 1951 getting my tuition through the GI Bill of Rights. I'd say there were about twelve of us, and no women.

—Joe Gagné, on earning a Toque at Le Cordon Bleu

Châteauroux's Golden Ghetto might have been an inviting way station for military nomads and their civilian supporters, but as we are about to see, there were others who set up camp in Berry who had been roaming for generations.

Hardly a tourist town, and never strongly considered a destination among sophisticated Frenchmen, Châteauroux was nonetheless a magnet for nomads, providing motivation that could eventually be either synergistic or discordant. Many were left wondering what had been real and what had been misplaced, wishful thinking.

Louis Gagné and his wife, Marie, left their mill at St. Cosme de Vair, France in 1644, heading for New France (Quebec). It would have been impossible for him to imagine that three centuries later, a descendant's odyssey would come full circle, ending at a Châteauroux hamburger joint.

In 1945, U.S. Army Mess Sergeant Joseph Gagné, found himself in post-war Paris wondering what the hell he would do next. The first tentative steps taken by the twenty year old son of an Augusta, Maine bakery owner were hardly auspicious, and for a guy who loved to drink, it could have been worse if not for a chance meeting with twenty-three year old Charlene Michot. She was hardly impressed by Joe at first glance.

"I met him after I had come out of the Metro [subway] and I was walking down the street. When I first saw him, I thought he was just another ordinary GI. It was not accepted for a woman to talk to a strange man on the street, but it was right after the war so it was acceptable to talk to GIs. It was also okay to invite them to your house," she said. Joe recalled that he was moderately successful in making his approach to Charlene, stumbling through the French he had learned as a kid in Maine. About two weeks later, Charlene, a born and bred Parisian working as a secretary, invited Joe to meet her parents.

"My parents did not think much of him. [Laughter from Charlene and Joe]. When he arrived, he was completely drunk! He was so completely drunk that my parents had to take him upstairs and put him to bed in my brother's bedroom.

"My parents came to accept him. They also accepted that he was a drinker. My mother and father got to know him because he spoke French."

It was not long after meeting her parents that the couple decided to get married. Then Staff Sergeant Gagné was transferred to Germany, returning to France a year later. He was unable to secure the documents needed for the couple to be married in France. As a result, they flew to Maine, got married, and it did not take long for them to decide that the "Pine Tree State" was not for them. The Cold War, as defined by Winston Churchill, began in 1947, and the young couple was among the early Cold War nomads.

It was the late 1940s, and after settling down in Paris, Joe was still wondering how he would make a living. Coming to his rescue was one of the most unlikely teaching venues accredited by the Pentagon, its approval coming in the midst of stormy debate in Congress and the highest American academic circles. Joe and Charlene could not have cared less about the vitriol, thankful that the contentious GI Bill of Rights would pay their bills while Joe was attending the world's most prestigious cooking school, Le Cordon Bleu.

The French were accustomed to an educational class system where it was *de rigueur* to attend the best schools, all the while knowing that the country's educational pantheon was closed to all but the privileged and most gifted. Appreciating this, the French could hardly be surprised at what was coming out of Washington and America's Ivory Towers. James Conant, President of Harvard University, and Robert M. Hutchins, President of the University of Chicago, believed that the GIs would be misguided and misplaced because most of them did not meet college academic standards. Edith Efron, a journalist writing for *The New York Times Magazine* in 1947, quoted one anonymous professor at Pennsylvania's prestigious Lehigh University, that ". . . men were being admitted to college who instead should be hod carriers." Another academic stalwart at Lehigh warned that the GI Bill was ". . . bringing about a 'nonsensical' situation in which every empty grocery store was soon to be converted into a 'college.' . . ." And then the empirical evidence started coming in. There was no reaction from Harvard President Conant to a 1947 *Life* magazine article, "GIs at Harvard: They Are The Best Students in College's History." *New York Times* Education Editor Benjamin Fine noted that, "The GIs are hogging the honor rolls and the deans' lists; they are walking away with the top marks in all their courses."

It would not be hard to imagine guys like Conant and Hutchins and the snobs at Lehigh University letting out a collective burp when they learned that Le Cordon Bleu was training and returning home a bunch of GIs who had traded in their hods and work boots for the most prestigious toques in the world.

"Just about all the students were former GIs. I was there in 1951," Gagné said. "I'd say there were about twelve of us, and no women."

With his toque in hand, Joe had to decide how to make the best use of it. So he headed south to Châteauroux and found himself among young GIs who, for the first time, knew that their dream of a once unattainable higher education was waiting for them, if they had the brains to pursue it. He would eventually become Châteauroux's most enduring American icon, one who would be honored at the 2007 reunion more like a rock star than someone who had dispensed French fries, chili, hot dogs, and hamburgers. Mayor Mayet, Senator Gerbaud, American pilgrims, and assembled French print and broadcast media paid homage to the eighty-five year old Gagné, who accepted the plaudits while seated slumped in the bright sunlight outside Châteauroux's exhibit hall. There are two divergent recollections that define the facts behind the myth that put Joe in that chair—first Joe's and then his wife's.

"I had heard that they were opening up a big American Air Force base in Châteauroux. I was looking for a job," Joe said. "I wanted to do something different. I came up with the idea for Joe's From Maine. The American base at that time had none of that stuff that GIs wanted. We gave it to them."

When the couple arrived in Châteauroux in January 1952, CHAD was already transforming the backwater community. "It was just like any other French town except that it had a lot of GIs. We were told that before they came, it was just a sleepy village without much going on. At seven o'clock, everything was closed and the town went to sleep," recalled the diminutive Charlene. She would eventually find herself cooking in high heels because she was too short to work at the restaurant's hot grill. "Joe was supposed to go to work at the base, but he didn't get the job. My father let us have a small house in town so we would have a place to live." The Gagné family, Joe and Charlene, their children, Annette and Mike, and Charlene's parents lived in an apartment above the restaurant. Helping out with the rent money during those early years was an old maid who lived upstairs. "She was a good tenant. She wasn't married, she didn't have children, and she didn't have GIs [visiting her]," Charlene said. "In fact, she didn't like American GIs. She viewed them as typical Americans, too noisy, staying up too late, things like that.

"After Joe didn't get the job at the base, many GIs asked him why he didn't open up a snack bar for them in town because there wasn't anything at that time in town for GIs. So we did. It was the GIs' idea, not ours, to open up a snack bar."

Working sixteen hour days while turning out two hundred to three hundred hamburgers, ham and cheese sandwiches, and BLTs, Joe and Charlene turned Joe's From Maine into a home away from home for young GIs, complete with a jukebox, and the basement Cave. "When things got out of hand, I had an electrical bell that was wired upstairs to the bar. It was loud and it usually quieted things down for awhile," Joe said. Business only got better when Joe and Charlene returned from a visit to Maine with a deep fryer, and the menu then included French fries, enough for them to hire a man who did nothing but peel potatoes from dawn to dusk. On paydays, customers lined up from the restaurant's front door on Rue de la Poste all the way down the block to Avenue de la Gare.

Joe Gagné, a life-long heavy smoker, died in late April 2009 at a Châteauroux hospital. He was admitted with a severe respiratory problem and was in intensive care for thirty days. At his request, his funeral was largely a family affair. News of his death went out immediately in the print and broadcast media, and the Internet took care of the rest. Emails bounced back and forth across the United States as thousands who had shared the CHAD Golden Ghetto experience mourned for the man who, along with his restaurant, were the symbols of what had been the happiest years of their lives.

They Flew the Coop

Robert Frenette's early military wanderings eventually took him to Châteauroux in 1960, six years after his aggressive abandonment at seventeen of what he described as a "chicken coop of a home" in Chippewa Falls, Wisconsin. "My family were like chickens, they were all over the place," he said. The Frenette home was one place a teenager with wanderlust could not get away from fast enough. "I ran away the first time when I was fifteen, but I really didn't run away because I told my parents I was leaving. I had no home. When you leave your home two weeks before your seventeenth birthday—and earlier, had left and returned for two weeks one year, and maybe one week another year— you really don't have a home." On the road a year later and no longer needing his father's consent, Frenette joined the Air Force at eighteen. When interviewed in 2005, the retired Air Force master sergeant was sixty-eight years old. He had been living with his wife, the former Sharon Bressin, in nearby Levroux, his wife's family hometown.

Monday is market day at Levroux, a town of less than four thousand people some twenty-five kilometers from Châteauroux. The market at Levroux, like countless others throughout France, is as much a social necessity as it is a mandatory shopping habit. There might be hundreds of men, women, and kids milling about the town square, but it would be impossible to mistake Frenette as anything but an American as he pedaled his bike among them on his way to Maurice and Chantal's Sports Bar. For Frenette, it was a morning ritual, as he parked his bike, made his obligatory *bonjours* and *ça vas* to regulars seated at sidewalk tables, and elbowed his way to the bar for his ten o'clock cognac. Frenette was wearing a hard-brimmed, baseball-style cap with a Green Bay Packers logo, and a sporty windbreaker that draped the shoulders of a man who had been accustomed to hard work his entire life.

During CHAD's golden years, there might have been hundreds of families scattered throughout Berry who would welcome an American airman into the fold, but the Bressins definitely were not one of them. Sharon worked at the base motor pool when she and Bob started dating in 1962.

"She worked at the base for ten years, but had never socialized with the GIs. She had no relationships with them. Her job at the base was timekeeper in the motor pool for two hundred drivers," Frenette said.

"Our marriage was a lightning stroke. I knew her for eight days when I asked her to marry me. She said yes. I never saw her family until about three months later. The family did not accept it. A lot of what was said was very personal, and I don't think I should talk about it. But it was about them not accepting it, when I asked them for permission.

"We went to the U.S. Embassy in Paris anyway to get the necessary papers. After eight months we still hadn't received the papers back. I went to Sharon's mother and asked, 'have you received the papers?' She always denied it. All that time she had been hiding the papers from the Embassy in the bottom drawer in the kitchen stove. She justified her actions by saying that our marriage would never last, not more than two or three years."

Judging on what she would see during the couple's first years of marriage at CHAD, Sharon's mother needed no justification for having stuffed Frenette's marriage application into the kitchen stove hiding place. Her son-in-law's insatiable restlessness soon surfaced. "We would work at the base from Monday to Friday, and on Friday night we'd leave and return to work on Monday. Sure, we had friends, travel, travel, Switzerland, Luxembourg, we honeymooned in Venice," said

Frenette, whose flight line duties provided him with the ample time needed to discover and rediscover European destinations.

It would be hard to find a better example than Frenette's early encounters with the Bressin family to illustrate the paradox that was Châteauroux's Golden Ghetto, which in many ways was a feast of contradictory impulses. Sharon, who had started work as a teenager at the base, had worked her way to the important, well-paying job of timekeeper for two hundred drivers. Frenette found it hard to understand why there was so much negative reaction by the Bressins, a family for whom the Golden Ghetto had provided very well. "Her father always worked at the base with us on the flight line, loading and unloading the planes. He worked at the base until 1967; he was one of the last to leave."

Besides the four years he spent at CHAD, Frenette pulled tours of duty in Morocco, Germany, Saudi Arabia, several bases in the United States, and Vietnam. His son was born in 1969 while he was stationed in Germany. Frenette balked when ordered back to Vietnam for a second tour, and with twenty years in the Air Force, he decided to get out. "I'm not a coward you know," he said. "I had already been there. I volunteered for three years. I wasn't in a hurry to go back and so I signed my discharge papers."

Upon listening to Frenette, it became obvious that during two decades in the Air Force he had been searching for the home he firmly believed had never existed in Wisconsin. Despite her family's early opposition, Sharon was providing her husband with a home that he had lacked since he was fifteen, and although she might not have realized it at that time, her husband certainly did. "I knew we were coming back to Châteauroux. It was never àny discussion. I know how the French people think. We just knew we were coming back. She would never make a home in the United States."

The Golden Ghetto disappeared seven years before the Frenettes returned to Levroux in 1974, its glitter inexorably fading into memory. Thousands of well-paying jobs had disappeared, with very few "Help Wanted" appeals in evidence. There were no fun-filled trips throughout Europe for Bob and Sharon, just hard times. His first job was at a Châteauroux foundry, and it did not last long. The job paid the equivalent of two dollars an hour, but even at that wage the competition was tough. Everyone was looking for jobs that did not exist. Frenette said he had been promised a promotion, and when he did not get it, he once again took to the road. Luckily, a friend was able to get him a job at a farm in St. Martin de Lamps. Finally, in 1981 luck changed for Frenette and his friend. "We had French friends who wanted to give up running the gas service station they had in Levroux. So we took it over, just the two of us, in 1981 and ran it for seventeen years."

A Fortuitous Missing Letter

The simple disappearance of one letter, the letter "T," made it possible for another family to begin a journey that would provide a new identity and eventual freedom from the racial fear and anxiety that had enveloped their native country. Châteauroux had never been the family's destination, but more than half a century of wandering through the thickets of European political turmoil, warfare, and economic collapse led it on an odyssey that would have made Homer proud. The decision to flee Germany did not come easy to the Dietz family, but how could it be otherwise when even the most politically naïve could recognize that the country was no longer a safe haven for Jews, if in fact it had ever been one.

I interviewed Lillianne Diez Herrera twice, the second time in March 2009. We learned earlier that Lillianne, as a fifteen-year-old, was headstrong enough to secretly gawk at U.S. airmen so she could bring back juicy gossip to the *lycée*, and a few years later she could laugh off jealousy-driven threats by young Frenchmen that she would have her head shaved as an American collaborator. She challenged her father's orders and continued dating GIs before her marriage to Air Force Sergeant Ernesto Herrera, and she paid the consequences. "My father slapped me in the face. It was the first and only time he ever hit me.

"Our family's original name was Dietz," Lillianne said. "My family simply removed the 'T' to make the change to Diez. My great-grandfather did that, not my grandfather. The name was changed when he—

my great-grandfather—got to France before the turn of the century. My family's Jewish background changed with the changing of our name, but we did not convert to Catholicism. But my mother had a Christian Bible, and you know how in the Bibles they used to put important dates, names and births? It shows clearly when they changed the name from Dietz to Diez. My brother Jean still has the Bible. It shows that they were Protestants, but my grandmother was Catholic, she came from a Catholic family."

The newly renamed Diez family nonetheless brought two strong, Dietz family traditions with them from Germany. One was a love for the pastoral life, particularly raising sheep, and the other was banking. On the surface, this appeared to be an odd mix, but during the first eighteen years of the twentieth century, the Diez family was capable enough to pull off this economic balancing act. The family brought their penchant for making money with them to France. "My grandfather had a lot of money; made a lot of money on the stock market. The Russians were building the Trans-Siberian Railroad, and he bought stock in it, a lot of stock! It was a really good deal. Then, of course, the Russian Revolution came. And then he lost almost all of his money, most of it," Lillianne said. She did not know how much of her family's fortune was invested and lost.

A stranger-than-fiction family history began to emerge, one with many permutations. Safely ensconced in the family haven near Longvic, not far from Dijon, it is doubtful that any of the family members had Châteauroux on their minds, or as newcomers to France, even knew it existed. The Diez clan would ride and fly their way into Berry, and in retrospect, it would only seem natural that there was a Golden Ghetto awaiting them.

Lillianne inherited her strong-willed stubbornness from her father, Jean Baptiste Pierre Diez, who set the pattern for family defiance. He had made his daughter pay a price with a slap in the face. Thirty years earlier he had paid a similar price, one that included a public humiliation. Training as an agricultural engineer, Diez faced the boring prospect of a lifetime raising sheep on the family farm near Longvic. He fled the family confines after deciding that the Foreign Legion was more to his liking. He would eventually find himself running a very profitable three-pump gas station for the U.S. military at CHAD. This followed an illustrious career as a French fighter pilot for *Armée de l'Air*, and later aas a Resistance leader in Berry and Normandy.

"He joined the Foreign Legion using another name, not his real name. I don't remember the name, but I remember him telling me that he took a Belgian name. He told the Foreign Legion that he was a Belgian, because after all, Belgium was right next to France. He had to assure them he was not wanted by the police; that he wasn't hiding out because he murdered somebody. After three years you could take your real name back, and the Foreign Legion gave you money. Because my father didn't have anything to worry about, he gave them his real name, Diez.

"The Foreign Legion had his real name, and at the same time, my grandfather was searching for him because he had left the farm. The Legion contacted my grandfather and sent my father back to him. That was either 1924 or 1925, when he was twenty-one.

"They were waiting for him when he arrived in Dijon. My grandfather and the other grandfather, his mom's dad, were waiting for him. You know, they were well dressed, the way they dressed in those days, with a top hat and all that, and those long coats, and each of them had a cane. When he got off the train at the station, they both started beating him up with their canes. That was right at the railroad station in Dijon." Despite the beating, Diez once again defied the family and this time joined the French Air Force.

The legend of Jean Baptiste Pierre Diez grew during the war years. Lillianne's brother, Jean, recalled: "We lived in Normandy at the time. We saw him about every two weeks. He was working under a false identity. As a member of the Resistance, he helped prepare for D-Day, work that assisted in the invasion."

Perhaps it was fate that Diez returned to Berry where, as a French Air Force fighter pilot, he flew out of the airstrip at La Martinerie. All evidence indicated that he had no problem making the adjustment from war hero to gas station operator, a job made possible by the American military with help he received from his high placed French friends in Paris. His gas station on the main road through Déols was in front of the former chateau of a French industrialist, a three-story building replete with terrazzo floors and marble. From 1961 to 1966, the chateau was the family home of U.S. Air Force Brigadier General Robert "Red" Forman, whose first command in France was as boss of the same Evreux-Fauville Air Base that Diez had commanded more than a decade earlier. Their journeys would place these two heroes only yards away from each other, one of them running a small gas station, and the other commanding the 1602nd Air Transport Wing.

Perhaps it was best that they had not met, since these two military men who shared a love of flying and adventure were so unalike when they had their feet on the ground. We will discover how the parties hosted by General Forman at the chateau only yards from the Diez gas pumps added to the Golden Ghetto legend.

Eight years earlier, Lillianne Diez was fifteen, and her brother, Jean, was twelve when the first gasoline started flowing from their father's pumps. Both of them acknowledged that their father liked the Americans up to a point, but their father's tipping point was reached when Lillianne started doing what came naturally for most teenage girls during the CHAD glory days. Listening to Lillianne, you could "fill in the blanks" when she described her relationship with her GI boyfriend. It was a secretive and seductive life that added glamour and adventure in a town where there had been only gray boredom before the Americans arrived. "When I was working at the Air Station, I met a guy I really liked. He was in the air police, his name was Jessie Phillips. I was still going to school at that time. He would pick me up at school and I would get into the back seat of his car and lay down on the floor so nobody could see me. It was a beautiful car, a convertible. We would drive out of town into the country and park on the side of the road. We would do some talking and we would kiss. It was innocent.

"I guess you could say it was because of my teenage vanity that my father found out about Jessie. I would go to the PX and buy lipstick, real French lipstick. I wasn't supposed to do that. I would put it on, and when I took it off you could still see that I had been using it. My father knew this.

"I knew it was only a matter of time before my father confronted me, and I feared when that day would come. One time, I was walking on the street when Jessie drove up and stopped. I leaned over and started talking to him in the car. My father saw that and when I got home he was furious, furious."

If there was a single plot line that emerged in dozens of interviews, it was that most French families in town or living in the rural country-side liked Americans, or at least harbored no animosity toward them, until it came to marriage. To get her away from Jessie Phillips, Lilli-anne's parents shipped her hundreds of kilometers south, to an air base near Aix-en-Provençe, to live with her older stepsister and her husband, General Germain d'Acrillé.

Jessie Phillips persisted, following Lillianne to Provençe, and back to Châteauroux. Diez told him to get lost, and if he wanted to marry his daughter to come back when she was twenty-one. Then came an arranged marriage to Antonine le Mount, a much older engineer from Paris. Two children, Antoine and Marilyn, were born. Both children now live in California. Antonine and Lillianne separated after eighteen months of marriage. Official divorce followed three years later. Lillianne was pursued by another American GI, Ernesto Herrera. At twenty-seven, she was no longer a kid. Her parents liked Ernesto and accepted him. The couple got married. They moved around a lot. There was an acrimonious separation. They got back together again. They settled in El Paso. Both became real estate bro-kers. Ernesto died of cancer before I could interview him. Lillianne retained her El Paso home until the spring of 2009, when, all alone—her son and daughter living more than a thousand miles away in Southern California—she pulled up stakes and moved to the high desert community of Temecula, California.

Meanwhile, her younger brother, Jean, had remained in Châteauroux. He was twelve when he started helping his father at the three-pump gas station. "That's how I got to know who the Americans really were," Jean said. "They were really friendly and very open in their dealings with the French. I was just a kid, but I still appreciated how generous they were.

"Working at my father's gas pumps was a real education. First, we didn't actually sell gas. In order to get gas you needed a gas coupon issued by the U.S. military. Generally, no questions were asked when the French presented the necessary coupon even though we knew the U.S. military did not issue coupons to the French. Even as a boy I realized there was a black market. These gas coupons were like gold."

Père Jean Baptiste Pierre Diez, once a Foreign Legionnaire, a much-honored combat flyer and French Resistance fighter, died of a massive heart attack while unloading groceries at his home, not far from Châteauroux. He was seventy-seven years old. At the time of his death, Lillianne was in America with her husband, Ernesto, selling moderately priced homes to blue-collar Mexican-Americans in El Paso, Texas. Her brother Jean, who never left France, was the owner and operator of a driving school in downtown Châteauroux.

LONELINESS TRANSFORMED

Barbara "Babs" Moody spent only her senior year attending CHAD High School, arriving late for classes in October 1955 and graduating the following June. During that nine months, Babs joined a select group of American military and civil service dependents, the "dorm rats." These were the kids whose parents were serving in military bases throughout central France not big enough or established enough to have high schools. By train or bus and often by both, they flowed into CHAD every Sunday night, returning home the following Friday night, adding yet another degree to their nomadic credentials. Babs fit in superbly, enabling her to break free of her cocoon of loneliness and anxiety. Within weeks, she was the uncrowned "queen of the dorm rats."

"She fit in perfectly from the very first day," fellow dorm rat Frank Nollette recalled. "She was beautiful and friendly to everyone, never had an unkind word to say."

"She was a real beauty. Just because of that alone, you couldn't miss her," said basketball and football star Dave Bankert. "She just took over, and with her looks it was natural that she became head cheerleader."

Her stepfather, Fred Lyons, was a tank commander during WWII. He was called back for the Korean War, served out his time and joined the Civil Service working for the Army. His assignment to the Army supply depot at Chinon, where there was no high school, provided Babs with what she called "the best year of my life." She seemed to be everywhere, taking advantage of every opportunity for acceptance by classmates, parlaying her easy smile and Marilyn Monroesque figure into prom queen, senior class officer, and head cheerleader status.

The nine months that this coal miner's daughter spent in the Golden Ghetto instilled in her a self-confidence that had been repeatedly suppressed by family and peers alike. For Babs, sports were the key that opened the social door that had been closed since her father's death.

"Sports were a big part of my life at CHAD. We were sent to a basketball tournament in Frankfurt, Germany. We were from the smallest school represented there. The teams were from the Air Force and the Army, which of course had some very big bases in Europe. We managed to win our first game and lost the second game by one point in the final thirty seconds. We surprised everyone by winning that first game. Compared to the other schools from the big bases, we were just a tiny little blot on the map.

"During the team's trip to Frankfurt, we were able to get away from our chaperone during the three or four days we were there. We ended up in a German beer hall, with all that oom-pa-pa music blasting away. There was plenty of beer flowing. The girls didn't drink very much, but the boys—you know boys. I believe they did some sampling. But those sausages, that was something else—delicious. We had a great time. We also went to a typical German restaurant and to a movie, which was in German, but so what, we enjoyed it anyway."

Babs could never have imagined the Frankfurt experience when she was attending high school in Uniontown, Pennsylvania. One of about fifty-five "dorm rats," Babs found herself willingly inducted into a peer group who's friendliness still amazed her fifty-two years later. "When I walked into the dormitory, I was overwhelmed by the friendliness, something I had never experienced before as an outsider coming into a new school. The next day when I went to class for the first time, it seemed like just about everyone came over and greeted me, wanting to be my friend.

"Every kid in that school had experienced the same kind of life. They had moved from base to base, from school to school, never staying long enough to develop normal friendships. They were constantly on the move, and when you have had that kind of experience you had tremendous empathy for others like you. I had not been associated with the military, but I had moved around quite a lot from school to school, and I knew what loneliness was. You generally came in as an outsider, these tight little cliques, especially among girls, were already in place. Any kid who experienced this kind of life knew just how tough it was."

Bab's father, Kerry Moody, was killed in an explosion at an Osage, West Virginia coal mine when she was only three and a half years old and her brother was one year old. "My mother was only nineteen. She raised us by herself and did not remarry until I was thirteen. Before my mother remarried, we were dirt poor, and I have to say we weren't that much better off after she married.

The CHAD dormitory that housed the dorm rats reflected middle American values straight out of an Andy Hardy movie. The dorm was a large Quonset divided equally between boys and girls, and romance blossomed between Babs and basketball star, Saul Wright, with the help of a kindly housemother.

"At night there was a man who monitored the boys on their side of the dorm, and a woman who saw that the girls behaved themselves. Our housemother was a woman about fifty years old named Mrs. Francel. As far as we were concerned her most important job was to pass love notes back and forth between the boys and girls. Separating the boys from the girls was a big military style bathroom and shower. The lights remained on all night. We would give Mrs. Francel our notes and she would make sure they would get over to the boys' side. There was no talking between boys and girls at night. She came on duty at eleven o'clock, and that's when it became fun.

"She would pass along the note and the only place you could read it was in the girls' bathroom or shower stalls. And you had to crouch down below the windows so no one would know that you weren't in bed. This went on every night. By the way, she also brought in wonderful strudel every night. She was much loved. I've often wondered whatever happened to her."

If one were asked to point out the transformative benefits of the CHAD Golden Ghetto, it would not be necessary to look any further than Babs Moody. When she arrived, she carried with her baggage that included an inferiority complex shaped by a childhood that included the tragic death of her father, poverty, and social exclusion. CHAD offered a quid pro quo, throwing out gift-wrapped opportunities and asking in return only that the wrapping be removed. Babs tore into the wrapping with both hands. She was described by a former classmate as "a real Elizabeth Taylor beauty," but she scoffed at that notion while producing swimsuit photos that supported her claim that she "was more the Marilyn Monroe type. The female figure was a lot different than today. That was probably the biggest attraction for the teenage boys. There was a lot of testosterone."

Shortly after her arrival she donned the head cheerleader's uniform, which would have been an unthinkable accomplishment at Uniontown High. Then followed three elections, Senior Queen, Senior Representative of the three other proms, and Secretary-Treasurer of the Teenage Club. The formula that did not work in Uniontown paid big dividends at CHAD, "I was friendly. I was not mean, and I made an effort to get along with everyone. I had friends in all the grades from freshmen to seniors."

After graduation, Babs spent a year "just hanging out" with a friend while staying with her family in a small home just outside Tours. When it came to loneliness, she was an expert. One incident that occurred during that year summed up the universal condition that prevailed among young GIs, no matter how good they had it.

"My girlfriend and I went into a brasserie not far from the Army base at Chinon. We walked in and we were the only girls in the place. After we sat down, we noticed a group of GIs sitting together at the other side of the restaurant. They had been looking us over from the time we walked in, talking among themselves and gesturing toward us.

"After a while, two of them walked over to us and asked where we lived and if they could walk us home. We found out that the GIs were the entire Chinon Army base baseball team. We agreed to go with them.

"We really didn't think we were taking any big chance. To say they walked us home would be understating what happened. They escorted us. They told us the team had run a little lottery to see who would approach us. Like us, they were just kids, and so very polite. At one point when we were outside, we decided to sit down and talk. They never made a pass and just wanted to talk.

"They wanted to know who we were, about our families, our hometowns. They were hungry for news from the States. They were probably in their late teens, possibly as old as twenty.

"They were just reaching out, wanting to be friends. Playing baseball, they probably had it pretty good, but underneath you could see they were just two lonely, homesick kids. It's been more than fifty years, but I still remember that evening. A good memory."

Upon her return to the States, Babs divorced her first husband after one year, remarried two years later to Harland Bennett, went back to college, never graduated, but concentrated on raising a family and building a business career. She has three children and five grandchildren.

The Golden Ghetto that Babs Moody left behind might have been "just a tiny little blot on the map" when she was there, but it was an ever expanding blot with no perceived boundaries, much like a Rorschach inkblot Test that allowed an observer's imagination to run wild, as long as Uncle Sam's dollars kept rolling in, and middle-class values formed the basis for reality.

From Name-Calling to Air Force One

The nomadic experiences of Sam Herrera and a young Anna Reh, which would eventually lead to their marriage at CHAD, could hardly be described as typical. Sam had pulled himself out of a southwest Colorado mine shaft and overcome racial discrimination in places he never expected, to forge an Air Force career of more than twenty years. Anna was only two and a half years old when she was carried on the back of an older sister as the Reh family successfully escaped from a Communist forced labor camp in Serbia.

Sam and Anna, ignoring the advice of his immediate superiors and chaplain, and opposition by Anna's friends, were married after a lengthy courtship. "At that time, most French didn't think this was a good idea," Anna said. "In fact, they thought it was a bad idea. They really looked down at you, that you were dating a GI. The worst reaction came from French men. They wouldn't talk to me, they ignored me. These were friends, people that I knew. I didn't let it bother me too much, you know, I didn't let it become personal. They called me names. Just name calling, nothing physical."

There were reams of paperwork demanded by the U.S. State Department and the Air Force that had to be completed before Sam and Anna could get married. There had to be extensive background checks to verify that she was not a Communist and that her identity as a Yugoslavian refugee was correct. Finally, on August 8, 1959, they were married in Saint-Chartier where Anna had attended school.

Anna was to learn that there were no boundaries for the stateside racial baggage that crossed the Atlantic to Europe, and even the maternity ward at the 7373rd AFH at CHAD was not off limits. She was convinced that her harsh treatment at the hospital where she gave birth to their son, Patrick, was motivated by her marriage to an enlisted Hispanic man. "When I was having my son at the base hospital, the nurse was really not very nice the way she treated me. Everything she did and said was very harsh. I was young, and that was kind of hard for me," she said.

A good portion of the more than twenty years which Herrera spent as a fuel specialist in the Air Force was with the Strategic Air Command, servicing bombers and intercontinental missiles. He retired as a chief master sergeant.

After his Air Force retirement, Herrera spent twenty-five years in the federal Civil Service doing what he did best as a fuel specialist—ensuring that no dangerous, impure fuel ever threatened the safety of the aircraft he serviced. His area of responsibility was New Mexico, and parts of Texas and Arizona. He was on call twenty-four hours a day, seven days a week, ever alert that the President was on his way aboard Air Force One [AF-1]. Then, with his retirement day rapidly approaching, Herrera became emboldened enough to seek a special favor that if granted, would forever define the end of his and Anna's nomadic journeys. He asked for clearance for them to tour AF-1.

His first request was refused. It would have allowed him aboard what eventually would be President George W. Bush's second-to-last visit to the southwest. "Finally, it happened. AF-1 was scheduled to visit New Mexico! I contacted my boss again to inform him of the ensuing visit, and he said he was going to do his very best to get me on board," Sam said. The approval eventually came through. By this time, Sam was considered a member of the family by the crew of AF-1, but when he asked if Anna could accompany him, his request was flatly rejected. However, Anna was allowed to attend a pre-visit briefing with Sam, and to walk side-by-side with him to AF-1 where a screening officer with a clipboard checked off approved visitors. He couldn't find Anna's name on the list and was about to turn her away when a ranking officer whispered in his ear and Anna's name was added. Once aboard, a guide took the visitors on an enchanting tour that Sam and Anna were attempting mightily to comprehend. "We continued on into the cockpit," Sam said. "There was a pilot or copilot there who asked me if I wanted to sit in the pilot's seat, and I said, 'Why sure.' As I sat there for a moment gazing at the controls, he asked me if I would like to taxi the plane for a little ways and my immediate answer was, 'No, not even an inch!' Of course we all laughed." Then there was the presidential executive suite, whose luxury paled when compared to Sam and Anna's opportunity to sit in the seats of power. "In the presidential office, I sat in President Bush's chair where his jacket was hung with his name on it. What a thrill it was to sit in his actual chair," Sam said. "My wife sat in Laura Bush's chair and I can still see her smile as we sat there for a moment!"

The odysseys of two nomads had come to an end, one that had begun when a teenage kid arose from his knees and left a narrow coal mine shaft in Colorado, and the other when a toddler escaped Communist tyranny on the back of her older sister.

1961 – COLD WAR POTPOURRI – 1961

- The Berlin Wall was erected by East Germany's communist rulers closing off the escape route that had previously been used by 3.5 million East Germans to escape to the West.

- From April 17–19th, in an attempt to overthrow the communist government of Fidel Castro, Cuban exiles trained by the United States in Honduras were defeated after going ashore at The Bay of Pigs in southern Cuba.

- On January 17th, outgoing President Dwight Eisenhower in his farewell speech to the nation warned of the danger presented by the growing power of America's "military industrial complex."

- America went Twist crazy, with Chubby Checker's "The Twist" leading the way for other twisters like Joey Dee and the Starliters' "Peppermint Twist," a nightclub favorite that became number one.

- In May, in Saigon, Vice President Lyndon Johnson called South Vietnamese President Diem the "Winston Churchill of Asia" just before 400 Green Berets were sent to prop up Diem's sagging army.

- On April 12th, the Soviet Union's Yuri Gagarin orbited the earth to become the first man in space, and on May 15th, U.S. astronaut Alan Shepard Jr. made his fifteen minute space flight.

CHAPTER 11

PATRIOTISM, BROKEN BOTTLES, AND PENICILLIN

*After Americans have lived abroad for a number of years they have lost
their usefulness as patriotic American citizens.*
—Major General William T. Hefley, defining military patriotism

When the C-54 Skymaster taxied into place at the Yalta airstrip on February 14, 1945, nosed down the runway and picked up speed for take-off, Lt. Colonel Francis J. Pope and five other pilots under his command circled overhead in the Crimean skies. The tripartite Yalta Conference had ended only a few days earlier. It would prove to be the last meeting between Russian Premier Josef Stalin, British Prime Minister Winston Churchill, and a very ill President Franklin Roosevelt, who died two months later. FDR had departed Yalta two days before. The famed 27th Fighter Squadron that Pope commanded had been chosen to provide escort for the Skymaster that was carrying Churchill and his staff. As it gained altitude and began banking southwest, Pope's six P-38 Lightnings took up their positions for an escort mission that would take them first toward Athens, then to Cairo for Churchill's last meeting with FDR.

Pope would command the 27th Squadron until war's end in Europe. Several other postings followed before he took over as base commander at CHAD during its death knell. If central casting at any Hollywood movie studio had been looking for a swashbuckling fighter pilot, Pope was their man. His hard-drinking, flamboyant, often-profane behavior added zest to a group of base commanders who had traded in their cockpits for procurement and supply balance sheets. Their careers began in the old "brown shoe" Army Air Corps, a world with rigid rules that became easier as an officer climbed the command structure. As was the case throughout the U.S. military, rank had its privileges, and Châteauroux's Golden Ghetto offered ample opportunity to take advantage of them.

Like Major General Joseph H. Hicks, the CHAD commander he succeeded, General Pearl H. Robey was near the end of his career when he took over command of the base. While Hicks was described as dour, terse, and critical, Robey seemed to be everybody's friend, including the kids at the base. "Whenever us kids were walking along the road and General Robey drove by in his staff car, he would stop and pick us up, then drop us off wherever we wanted," said John Natt, a student at the base high school and son of Colonel Theodore M. Natt. "General Robey was just a nice guy, a lot different than General Smith, who would never stop and just had his staff car drive by as if nobody was there."

Major General George F. Smith might have been disdainful when it came to youthful American hitchhikers, but he had unbridled enthusiasm for his first love, golf. It is impossible to discern whether John Natt's criticism of Smith was his own or was adopted at the family dinner table. We learned earlier that Smith had decided that if there was to be a golf course on the base, then by golly it would be golf at its democratic best. Every Saturday, Colonel Natt and every other officer not on duty, would stoop shoulder to shoulder with the lowliest enlisted man, clearing rocks, stones, and weeds from the newly created fairways. There was nothing too good for Smith's golf course. A professional greenskeeper was imported from Scotland at U.S. taxpayers' expense, and became a permanent member of the golf course staff. Smith apparently did not want his beloved fairways pock-marked with divots left by untutored golfers, so he imported golf pro Ted Wilson from England to give lessons. But first, an indoor golf practice area, complete with protective netting, had to be installed in the base gym. Wilson appreciated that a well-paying teaching gig had been laid in his lap by Uncle Sam, and as a natural charmer, he also knew that the throng of military wives taking lessons had to be happy. There was a photo taken at an off-base officer's party with Wilson draping his arm around a smiling Major Donna Hildebrand, in civilian clothes, each with a drink in hand.

The command officers at CHAD were experts in procurement and supply, and as was the case with General Smith's beloved golf course, the cynical truism "taxpayers' money has no conscience" certainly applied. It was easy to lose track of why CHAD was there in the first place, that without the Cold War and NATO there would have been no Golden Ghetto. No one could have been more concise in describing NATO's mission than Lord Hastings Lionel Ismay—"Pug" to his friends and colleagues—who was appointed NATO's first Secretary General in 1952. He explained that it was "to keep the Russians out, the Americans in, and the Germans down."

De Gaulle was never happy with a NATO dominated by America. De Gaulle left no doubt in anyone's mind when he declared on November 23, 1959 in Strasbourg, Germany, "Yes, it is Europe, from the Atlantic to the Urals, it is Europe, it is the whole of Europe that will decide the destiny of the world." There was no mention of the United States being a partner. It was also noted that "none too subtle" imperial warnings such as this one were largely ignored in Châteauroux where the good times kept rolling on thanks to American dollars.

Then, on March 7, 1966, Charles de Gaulle delivered his *coup de grace*, announcing that he would pull all French forces out of NATO, boot the Americans out, and that NATO headquarters was no longer welcome in France. The headquarters was moved to Belgium.

Simone Nickles had worked for NATO for more than ten years, first in northern France, and then in Châteauroux. Fluent in French, German, Italian, and English, she was a valuable asset and NATO wanted to relocate her at its new headquarters, but she refused because she was romantically involved with the man she would eventually marry. She considered De Gaulle's decision insufferably devoid of any real feelings for the people it would affect. "I was very mad. I had made many friends here. I was friends with American wives and many others.

"At CHAD there was widespread disbelief and sadness among the NATO people. Many Americans had married Frenchwomen. It was like one big family. These people were happy. De Gaulle's decision killed a great deal of their happiness.

"We learned the news when our NATO boss took us all into a big room and said, 'I am sorry to have to tell you about the decision made by Charles de Gaulle.' You could see that he was emotionally distressed. He did not say too much, only that he was returning home."

Simone noted earlier that she was paid the best wages of her life and, as a NATO employee, had special privileges not available to French employees working for the Americans. She could use the PX, the commissary, all other base facilities, including the golf course, was issued coupons for fuel purchases at the Diez gas station, dining at the Officers' Club, and attending dances and special shows at the club. It was truly a good life.

GOING NATIVE

When Major General William T. Hefley arrived in May 1958 to take up tenancy at CHAD in order to consolidate procurement and supply operations in Europe, he brought with him a résumé that would make your eyes glaze over if you studied it in one sitting. As the world would discover, his credentials did not include a certificate in how to avoid offensive and ultimately damaging hyperbole when talking to the press. Hefley was quoted at the beginning of this chapter defining one of his iron-clad rules for American patriotism. His declaration came in an article by Bill Russell in the American Weekend newspaper supplement in February 1960. It caught the attention of Art Buchwald, and therein began a public relations nightmare for Air Force brass around the world.

With a name like William Tell Hefley, it was as though he had long ago been fostered by Gioachino Rossini to be his son in the *William Tell* opera. With a homegrown apple firmly in place on Hefley's head, the word was out that the press could un-sling their bows and shoot at will until their quivers were empty. Guess what? They did exactly that. With quotes like this from Hefley, their target could not have been easier. "Long stays overseas affect an American's patriotism and loyalty to his country. Most of the people who don't want to go back home don't have the slightest interest in the U.S. except for the high American pay they receive." This prompted Buchwald to observe, "You have to admit, this interview was a beaut."

And it did not end there as Hefley, who wanted all tours of duty for the American military and civilians alike to be limited to four years, said: "If an American wants to go native, it's hard to see why they don't take out citizenship in the country that they prefer to their home country. If an American knows the local language, he may be injecting wrong ideas into his dealings with the natives."

Buchwald noted that Hefley's statements "came on the heels of the discovery that the U.S. Air Force was putting out manuals on how to walk a dog, buy whiskey, and some which warned NCO's to beware of the National Council of Churches because there were many card-holding Communists in the pulpits—it's bound to make people wonder just how much hot air the Air Force is producing on the ground as compared to what it is producing in the sky."

Much to the consternation of Hefley and his superiors on both sides of the Atlantic, Buchwald spotted this as an opportunity to unleash his special brand of satire. "We were fortunate in interviewing the other day Major General Roger Wilco-Out of the First Airborne Foot-in-Your-Mouth Training division. Besides passing out all manuals concerned with such military subjects as 'How to Put Shoes on in the Morning,' 'Sports Car Recognition,' and 'Learning to Brush Your Teeth,' General Wilco-Out is also concerned with training general officers for public relations."

Buchwald was unsparing in describing a mock interview with General Wilco-Out, asking him how he felt about Hefley's comments.

"'Excellent, excellent. We're all very proud of General Hefley and we're putting him in for the DFC, the Distinguished Foot Citation.'

'Then you agree with what he said?'

'One hundred percent,' General Wilco-Out said. 'You can't trust an American when he stays abroad for any length of time. Gets contaminated meeting all those natives and everything. First thing you know, he learns a language—very bad for an American to learn a foreign language. Show me a man who speaks a foreign tongue and I'll show you a man who's selling atomic secrets to the Russians. In spite of great generals such as Hefley, there are still some Air Force officers fraternizing with Frenchmen, but we're going to find out who they are and send them home.'

"'General Wilco-Out, has the Foot-in-Your-Mouth Training division done anything to help the situation?'

'Well, we've just put out a new manual for our personnel entitled "Speak English or Die." There are some top secret projects I can't talk about right now, other than to say we think there's a subversive link between Americans who live overseas, the National Council of Churches and the United States Navy. And if we can prove it, we're going to blow our entire security system sky high.'

"'Well, thank you, sir. *Merci beaucoup.*'

"General Wilco-Out looked up startled, 'What was that you said?'

"'I said "thank you very much" in French.'

"'I thought so,' he shouted. 'MP, arrest this man!'"

Best-selling author and columnist, Robert Ruark, took umbrage at Hefley's comment that Americans who had lived for more than four years overseas were "homesteaders" and might as well stay overseas. Ruark wrote, "I particularly like the phrase 'wants to go native.' It connotes a delightful picture of a bearded man surrounded by palm trees, dusky children, exotic women, pigs, goats, and cheap local brandy."

Hefley's superiors in Wiesbaden, Germany quickly voiced their displeasure, and an explanation, if not an apology, followed almost immediately. "At no time during my interview with the press did I question the loyalty of the vast body of Americans serving with the armed forces overseas. My statement was directed only to that small number of Americans who don't want to ever return to the United States. I further stated that long stays overseas do not affect all persons in the same degree."

Until his fall from grace, Hefley had an unblemished Army and Air Force record. When Hefley returned to the States in 1960, he left in his wake a legion of disgruntled Americans forced to return to the States after four years of service in the European-African-Middle Eastern theater. Nick Loverich was among them, leaving his wife, Françoise, and young daughter behind in Issoudun while he toiled over the paperwork necessary to get his old job back and rejoin them in France. An accomplished graphic artist, he would eventually return, and worked at CHAD until its closure in 1967. "I really have some bad feelings toward General Hefley," Loverich said. "There was no reason why I had to return to the States, and it was only because Hefley mistakenly said that anyone who stayed longer than four years was unpatriotic and not a good citizen. He was wrong."

Loverich's distaste for Hefley was shared by Ruark, who noted in his column that during his four years in Europe, the last two in France, Hefley had been living it up in Paris as "Head of the Air Material Command, another name for being chief storekeeper." During this time he enjoyed a "lavish living allowance" that enabled him to savor "truffles, *paté de fois*, and pressed duck, . . . *truite almondine* and cherries jubilee." If Hefley was a "storekeeper," the millions of dollars he saved Uncle Sam each year attested to the fact that he was a very good one, and this earned him a Commendation Ribbon, which mentions his ". . . unusual ability as an administrator" which enabled him "to obtain maximum production efficiency."

THE COLONEL'S FIREPLACE

If Hefley advanced his Army and Air Force careers in a world of cost and efficiency charts, time and motion studies, and equipment repair or replacement costs, it was a world that seemingly had no room for a maverick like Colonel Francis J. Pope, who brought a fighter pilot's mentality to his job as CHAD commander during the base's final years. In my hundreds of hours of interviews for this book, I never encountered as many rolling eyes and "it's hard to believe" shrugs than I did when the discussions centered on Pope. October 1940 is as good a place to start as any. Pope had been a flying cadet at Randolph Field, Texas for less than four months when it struck him that the Army Air Corps was cheating him out of the pay he deserved. He was drawing twenty-one dollars a month in private's pay, and was firmly convinced that he should be getting seventy-five a month as a flying cadet. Pope's superiors felt he did not have enough flying time to be paid the larger amount, but he persisted. He finally wore them down and they agreed to pay him seventy-five dollars in back pay, and the same amount from that point on. They still doubted he was entitled to it, but gave it to him nonetheless.

After getting his wings, Pope eventually found himself in the cockpit of the twin-engine P-38 Lightning, one of America's hottest fighter planes during WWII. The Lightning could easily cruise at 360 miles per hour and had a destructive arsenal of .50 caliber machine guns. A modified version, the P-38F, had two drop fuel tanks and was an ideal fighting machine for the vast distances protected by the Alaska Defense Command. As a member of the 54th Fighter Squadron, Pope flew seventy-five combat missions against the Japanese in the Aleutians. The historian for the 54th Squadron describes one memorable mission when Captain Pope and three other pilots "hit the jackpot," during a low-level mission to Attu when they destroyed eight Japanese "Rufe" float planes, as well as billets for the support troops.

Pope would eventually take command of the 27th Fighter Squadron, the oldest and most renowned in the Fifteenth Army Air Force based in Italy. A 1945 cartoon appearing in Pope's hometown paper in Oakland, California shows a stern-faced Pope with a mustache, wearing the crushed and battered hat he had earned after flying twenty-five combat missions. The cartoon described Pope as a hands-on squadron commander, stating, "When Pope makes the rounds of his Squadron, anything can happen—he often operates the phone switchboard!" The acclaim earned by Pope and his 27th Squadron made them an easy choice for the Yalta escort mission described earlier.

When Pope arrived at CHAD to take over as base commander, it was with the belief that, despite it being a supply depot, it was still an air base, and an air base had pilots, and pilots flew airplanes. The first of two notorious incidents at the officers' club indicated that it was hard for him to accept that the bulk of the pilots at CHAD were desk-bound and only flew enough to keep their ratings. Captain Marty Whalen, reluctantly taken from his C-118 pilot seat to run the financially troubled club, described it this way: "About a year before the base closed we had a new base commander, Colonel Pope. He was a fighter pilot from WWII, and had his own ideas about a lot of things going back to WWII. One day he walked into the club and asked, 'Marty, where the hell is the fireplace? You need a fireplace. Where the hell do you throw the glasses?'

"I could see that he meant what he said, and I told him 'We just don't do things like that here.' About a week or so later, the officer in charge of base building projects came to the club and wanted to know where to put the fireplace. I should explain that during WWII it was a custom to finish your drink after a mission, and then break the glass in the fireplace.

"It was then I bowed to the inevitable. I knew the fireplace had to be built, but not inside where people often walked around in their stocking feet. I could just see glass and blood all over. That wasn't the end of it. Colonel Pope decided there would be a big fireplace coming-out ceremony, and everybody had to wear flight suits.

"This led to yet another problem. Most of the officers at the base flew only enough to keep their ratings. Most of them had desk jobs and didn't have flight suits, so about a hundred new flight suits were ordered and issued out, even some French pilots got them. And at the same time, I was looking for ways to hide the glasses from the bar. If more than a hundred of them were smashed in the fireplace, there would be hardly any left and I would be out of business. So I had the bartenders and service help round up as many soda bottles, beer bottles as they could, then give them to the pilots to put in the pockets of their flight suits.

"When the big night came, Colonel Pope started things off by throwing back his drink, walking out to the patio and smashing his glass in the fireplace. He was followed by a hundred guys in flight suits, most of them very reluctant, smashing their soda and beer bottles, and some glasses from the bar. Then they went back into the club and it was all over. After Pope smashed his glass, he just kept walking without turning around, got in his staff car and drove off. He never knew the other pilots were breaking soda and beer bottles. That was it. As I recall, it never happened again."

When it came to opinions of base commanders, Katherine "Kay" Wood could easily be described as an expert. A graduate of Eastern Washington State University, Kay was a Civil Service recreation specialist who had seen commanders come and go by the time she reached Châteauroux in 1957. During the previous ten years, she operated Service Clubs at air bases in Japan; Rapid City, South Dakota; California; and Etienne, France, but had never met anyone quite like Colonel Pope.

"I ran into him at parties. We even had him over to our house for parties. He behaved just like a drunk fighter pilot, but I guess he was all right—just wasn't my kind. He wasn't tall and he was kind of wiry. I'm trying to think of who I could compare him to. I guess you could say kind of like Hollywood movie star Alan Ladd.

"I know my opinion of Colonel Pope was shared by others. I do know people who had to put up with him on a daily basis, and it wasn't easy because he was such a strange, loud, and profane person. His secretary was a friend of mine, and she just about stood on her head to get herself the hell out of there."

Kay was eighty-four years old when interviewed, and as we will discover later, retained an uncanny degree of recollection of her years at CHAD. But she could not remember whether Pope had ever made use of her Service Club, where the sedate activities included woodworking, photography, and trophy engraving for the various base athletic teams. According to her, he favored the much more boisterous Officers' Club. "He really liked Happy Hour at the Officers' Club. I don't think he ever missed one. He was very loud when he was drinking, and he was drinking all the time. I don't know if he drank while at work, but he drank at the club," Kay said.

"One of the classic things involving Colonel Pope that I remember was when a bowling game was set up at the Officers' Club. He had them line up wine bottles, beer bottles and glasses, and knock them over. There was glass all over the place."

As the inevitable base-closing drew nearer, and despite his eccentricities, Colonel Pope might have been the best man for the thankless job of overseeing the diminishing glitter of the CHAD Golden Ghetto. Just as Major General Joseph H. Hicks was probably the best bet to get CHAD operational sixteen years earlier.

THE BLACK MARKET

The presumption that American military bases overseas provided foster homes for black marketeers is certainly true at Châteauroux. But to what degree? Few Berrichon who admitted their involvement conveyed a sense of embarrassment or guilt, and from their vantage point, why should they? Hadn't Uncle Sam arrived in order to provide opportunities? So with a sticky finger here and there, a GI boyfriend in the right place, or access to the goodies at the base, why not take advantage of them?

"There were plenty of opportunities to make a buck, black market for example. Cigarettes and whiskey, as well as blue jeans. I can't remember how many pairs of blue jeans I bought at PX prices and resold in the villages nearby, making a dollar or two on each pair. So I made a buck [laughter]," recalled Annette Gagné while taking a break from the kitchen at Joe's From Maine.

"I remember everybody took advantage of the U.S. presence. When they decided to extend the runway, a local company was supposed to bring raw material for the concrete by trucks," said James Bessett, who earlier recalled he and his companions had no trouble spending the money that fell from the pocket of a drunk GI at Le Lido. "It was said that one of every two trucks that came into the base full of material, left full of material, too. The trucks were not unloaded each time. The company made a lot of money."

"My manager was fired for black-marketing cigarettes," said Madelaine Dagot, a clerk who succeeded her boss at the La Martinerie PX. "I recall him telling me at inventory to total the boxes of cigarettes even though they were empty. Cigarettes and whiskey were two of the most popular black market items."

"There were a lot of young girls in Châteauroux. They liked chocolates, cigarettes, and nylons. So, are you getting the picture? You might say there was a lot of testosterone flowing," said Dave Bankert. "And it was cheap; you could buy chocolates and cigarettes for nine cents a pack and trade them in town."

"The black market was big. Coupons for gasoline, everybody wanted them, cigarettes, whiskey," said Alain Birckel. "I worked out a deal with my friend Danny at the base. One of my best customers was Monsieur Rafa, the owner of the best men's clothing shop in Châteauroux. He was a Jew and a Zionist. It was usually two cases of cigarettes and two cases of whiskey. I understood he didn't do this for himself. He would sell them for a profit and the money would go to Israel for the Zionist cause."

Bob Goggin, a JAG attorney at CHAD, recalled a one-man black market operation which, while humorous, at the same time indicated how widespread the black market was. Goggin was ordered from Châteauroux to defend an accused airman. "It involved an Air Force officer who was staying at a Paris hotel and when he went to bed, he realized he was sleeping on U.S. Government issue sheets. It turned out the sheets came from one of our air bases in France. An airman at the base, he was either a sergeant or a master sergeant, was the quartermaster who handled the linens. He decided he was going to help the French economy by supplying hotels with sheets and linens at a cut-rate price. I defended the guy. He was busted in rank, but remained in the Air Force."

"With what the Americans had to offer, it was not a problem for us," said Gypsy Marien, the operator of a downtown Châteauroux bar who insisted his "hostesses" were not "working girls." "It would take place in bars. We would make deals for the tax-free gas vouchers the GIs would get. There was business concerning whiskey, cigarettes, and gas. A guy would also bring a pair of jeans or a shirt from the PX. That's when we would exchange dollars. Americans were never paid in francs."

"The black market was extensive. Every American had at least a few things he could dump off to make a buck, but not everybody did it. There were some people who specialized in it," said René Coté. He noted that as the French alcoholic palette became more sophisticated, their tastes ran up the connoisseur's chart from rye to bourbon, and finally to upper-crust scotch.

Black market profiteering involving U.S. military personnel in Europe had been around for quite awhile. It has reached such proportions that as early as August 1946, Colonel Francis P. Miller, an Army Intelligence officer, told a congressional committee in Washington that he was "morally shocked" by the widespread black market activities in Germany going back to 1945, which clearly pointed to the "moral disintegration" of all military ranks.

Predictably, congressional investigations followed revelations such as these. As a political firestorm engulfed Washington, the U.S. military tried to repair what had become a serious public relations disaster. By comparison, the black marketeering going on in Châteauroux during the Cold War could be called bush league, with military brats exchanging chocolates and nylons in the hopes of female favors, blue jeans pulling in a big two bucks profit per pair, and with illegally obtained gasoline coupons used to fill the tanks of a *Deux Chevaux*.

The Cold War and the bundling of the United States with eleven other nations into NATO was something new for Uncle Sam. The burgeoning black market in post-war Europe was conducted in comparative peace and tranquility, and, if you were to believe the leniency of penalties, it received de facto acceptance by the authorities.

When Josh Getzoff received his captain's bars shortly after his arrival at CHAD and began drilling and pulling teeth at the base dental clinics, he had only one thing in mind when it came to his off-duty hours: have as much fun as possible. He loved Paris, virtually camping there on weekends, often with special service officer Godfrey Friehofer Russman, Jr., one of his two roommates in Châteauroux. It did not take long for him to discover the magnetic pull of that city's black market. "But there's one thing I want to tell you that I'm not too proud of," Getzoff said.

"I went to Paris with Russman and he told me, 'Go buy all the cigarettes and Johnny Walker that you can, and other stuff like that from the PX.' So we drove into this place in Paris, each with a suitcase of clothes, and another stuffed with things from the PX.

"It was a prearranged meeting place. I was sitting in the backseat of the car when this guy, a Frenchman, got into the front seat and we started driving around Paris. We stopped and he goes around to the trunk and takes the suitcase out with the stuff from the PX and then he disappears. Here's what was going on. You could make enough money in Paris to comp your entire weekend without paying a cent, just by selling the stuff from the PX. They went kooky over the Johnny Walker and you could really get top price for that.

"Then later I got to know of other people in Paris who were a lot seamier. This guy approached me one time. He was an American expatriot and he said to me, 'I know you're a doctor. Can you get me penicillin? I'll pay you really good for it.' I can't remember what the price was, but I will tell you it was an absurd amount of money. I told him, 'Look man, I don't do any kind of drugs.' What they did in those days was to water the penicillin down and then sell it. I've heard that they even used soda water to cut it down, so ten units of penicillin would become a hundred units. I don't know how widespread this was at the 7373rd AFH. There was one guy from Châteauroux, and he was caught and shipped back to the States, probably to Leavenworth. He's the only one I know of."

Eleven years before Getzoff's arrival at CHAD, deaths caused by the underground sale of penicillin was common knowledge, enough to form the basis for a Graham Greene novel, *The Third Man*. By the late 1950s and early 1960s, black marketeering in and around Châteauroux was banal by comparison, with blue jeans, cigarettes, booze, gas coupons, and possibly frozen chickens and turkeys providing the way for GIs and the French to make a fast buck or two, posing the question as to whether or not the Golden Ghetto represented progress.

1962 – COLD WAR POTPOURRI – 1962

- On August 22nd, French President Charles de Gaulle escaped in his bullet proof car an assassination organized by French army officers who believed they had been betrayed by de Gaulle's Algerian policies.

- On August 5th, movie star Marilyn Monroe was found dead at 36 in her Brentwood home. Death was blamed on barbiturate poisoning and ruled a "possible suicide," leaving many to suspect a cover-up.

- On May 18th, Soviet Premier Nikita Khrushchev predicted, "The French fought 9 years in Vietnam and were kicked out. The Americans may fight fifteen years if they want to, but it will not help."

- On January 1st, the Rose Bowl game on NBC was the first coast-to-coast, color television broadcast of a college football game, with Minnesota defeating UCLA 21 to 3.

- On February 20th, Astronaut John Glenn Jr. attained hero status, becoming the first American to orbit earth, circling the planet three times in less than 5 hours, and then splashing down in the Atlantic.

- In October, nuclear war was averted during the Cuban Missile Crisis when U.S. spy planes discovered Soviet missile sites, later dismantled after a U.N. brokered agreement between the U.S. and Soviets.

CHAPTER 12

BETTY BOOP AND THE WING WALKER

If you want to see just a perfect little she,
wait till you get a view of Sweet Betty.
Boop-oop-a-doop.
—Song describing cartoon character Betty Boop

A cartoon that appeared in *The American Daily*, a newspaper circulated at NATO air bases throughout Europe during the 1950s, would undoubtedly have elicited at least a smile from cartoon satirist, Al Capp. His cartoon strip, Li'l Abner, lampooned the gaseous pretensions of politicians and assorted self-important blowhards while transforming the title character and his Dogpatch cohorts into a daily "must read" in newspapers throughout the country. One of the strip's stalwarts was Senator Jack S. Phogbound, a righteous, potbellied, filibustering "good ol' boy" who insisted his taxpayer-funded junkets made the world safe for capitalism.

Châteauroux drew the attention of acid-penned cartoonist Jake Schuffert, whose notorious latrine humor delighted GIs while driving more than a few base commanders to red faced apoplexy. One of them even had his cartoons banned from the base. In this case, a Schuffert character paid sarcastic homage to Al Capp's Senator Phogbound. Stepping from an Air Force transport, a fat, rumpled, swollen-nosed politician addresses assembled reporters at the CHAD air strip with a statement that displays both his bumptiousness and ignorance: "I'm going to investigate the high living of the troops overseas. Why, I even hear some of them are living in a chateau at Roux."

While there were certainly chateaus available to Americans in and around Châteauroux, two that will be discussed at length in this chapter, it was basically middle class living accommodations made available by the military that made American airmen and their families feel they had never left the United States. This was an era of innocent, boy-meets-girl romances, unquestioned parental wisdom, fair play, and unlimited horizons if you played by the Golden Rule. And for the kids, these rock solid assurances were supported by a print and movie cartoon industry that extolled simplistic fun even when the storylines were violent. The good guys always won.

When cartoon pioneer Max Fleischer created Betty Boop in 1930, he had no idea that he had given birth to a character that would become the most enduring sex symbol in cartoon history, still admired and lusted after in the twenty-first century. Betty Boop became synonymous with sex, with her wide seductive eyes, button nose, hoop earrings, tempting cleavage, short dresses, high heels, and garter belt. With all this prurient allure, Betty Boop nonetheless was able to remain innocent, in one movie cartoon fighting off an evil circus ringmaster, and then proclaiming, "No, he couldn't take my boop-oop-a-doop away!"

Boopy Forman was the incarnation of Betty Boop. Her husband, Brigadier General Robert "Red" Forman—with a chest full of medals, and genuine hero status—was the catalyst that provided the chemistry for a compound composed of diverse elements. There was one constant; an admiration for Forman that bordered on hero worship coupled with a deep fondness for Boopy. The officers at CHAD mixed easily with the town's rich elite, pursued their pleasures in mansions and on golf courses and ski trips, while having at their disposal a handy, four-engine transport. Surprisingly, members spoke openly and with candor when describing escapades that defined officer class entitlement, with little concern about consequences. Their stories told just how bright the glitter of the Golden Ghetto could be when privilege applied the polish.

When General Forman took over command of the 1602nd Air Transport Wing on September 29, 1961, including the 322nd Air Division, he and Boopy also laid the foundation for a clique that existed in parallel universes. In one orbit, its members performed their military and civil service duties with pride and diligence; and in the other, they raised fun and games to a level that would make denizens of the legendary "Animal House" envious. Forman was fifty-one; Boopy was thirty-eight. Call it magnetism or charisma, they were a team that attracted much younger men and women. "First of all, General

Forman was a unique guy," said John Riddle, who would retire as a major general in the Air Force Reserve. "Younger people under his command gravitated to him and Boopy. His energy was boundless. He didn't sleep a lot, didn't need much sleep. He was always up and around the base checking on things, and as a result he knew very much what was going on."

Shortly after arrival at CHAD, there was an incident that made clear to Riddle how and why the Red Forman legend had grown over the years. "The Communist threat was at its height, and I remember one incident in particular at the base that could have become very ugly. General Forman and I were sitting in the back of his light blue staff car, going between the air base in Déols and La Martinerie. When we came to the main gate at La Martinerie, there was a crowd of several hundred Communists blocking the way, shouting slogans. The crowd was so large that it spilled out into the street, making it impossible to drive to the gate.

"There were plenty of anti-American placards, a lot was going on. Because we had come to a standstill, they surrounded the car and began rocking it. General Forman kept his cool. He ordered the driver to keep driving, slow enough so the Communist protestors could get out of the way, but making it clear that we were going to keep going. And that's what we did, finally making it through the gate."

Jay Parsons, who succeeded Riddle as Forman's aide, said, "Red was the most energetic man I have ever known. He was no longer a spring chicken, but he still took stairs two at a time and was always on the go. I asked Red if he would ever slow down and he said, 'after you've been dead for ten minutes you will have all the time you need to catch up on your sleep.'"

Major Donna Hildebrand and her husband, Colonel Floyd C. Hildebrand—"Hildy"—transformed the former hunting lodge of Ferdinand de Lesseps in the village of Meunet Planches into a legendary fun-house while they were stationed at CHAD. Donna never put her chemistry degree from the University of Colorado to use in the Air Force, instead settling in as a special services officer. She and Floyd fit naturally into Red and Boopy's cadre. "Red was kind of gruff and outspoken, but he was always a perfect gentleman to me. For me, he was always a kind of polite, Southern country gentleman. But I know that if he got angry, someone could really get run over."

Dozens of photos chronicle what, to an outside observer, appeared to have been six years of party with interludes of work fitted in so everyone could catch their breath. Boopy was center stage in many of them, most of the time with a big smile on her face and a drink in her hand. She fit in very well with Red, a former wing walking pilot on the carnival barnstorming circuit of the late 1930s. There was also a stint ferrying Lucky Luciano's prostitutes back and forth between Memphis and Chicago, before Red was coaxed into the Army Air Corps by then Captain William Tunner, who would eventually command the Military Air Transport Service (MATS) as a Lieutenant General. "Red and I were not the typical Air Force West Point couple. I was probably the worst kind of an officer's wife," Boopy said. "Because of Red's rank and position at the base, I was usually the President of the Officers' Wives Club and I wasn't good at it. In fact, I was bad at it. Fortunately, the wife of the base commander took over.

"I really can't say how much friction there was between me and the other officers' wives. After all, this wouldn't be something they would talk about openly. I was the general's wife and I didn't give a damn. But yes, I'm sure they thought that I was awful. You'd have to talk to one of them to find out.

"So I didn't get deeply involved in any of the usual officers' wives activities. I took up golf and skiing when I was forty years old. So in the winter we would go to Austria and Switzerland to ski, and in summer I virtually lived on the golf course. So as you can see, you can't really write anything about me being a 'good officer's wife.'"

This seemed to make no difference to the members of the inner circle. "She was a neat, neat lady. She was petite and very friendly. She had that special kind of smile that made her eyes sparkle and her face really light up," Riddle recalled. "She was extremely clever, a great person to be around." Boopy's sense of humor could have an edge to it. Parsons remembered an incident at the Officers' Club: "I saw that her glass was empty and I asked her if she wanted another drink. She looked me in the eye and said, 'Don't you ever ask me if I want another drink. You ask me if I want a drink.'"

The socially acute Formans quickly become involved in Châteauroux's active, upper-crust party scene. Much of the activity revolved around the Balsan family chateau in Châteauroux. The Balsan family, whose fortune came from almost a century of fabrics and textile manufacturing, knew how to party. Etienne Balsan was one of Coco Chanel's lovers; Lt. Colonel Jacques Louis Balsan married Consuelo Vanderbilt on the rebound after her divorce from Charles Spencer-Churchill, Ninth Duke of Marlborough, who Consuelo's daddy had paid 2.5 million dollars in an arranged marriage. It seemed like everyone in the Balsan family loved to travel, loved Paris, and found it hard to turn their back on a good time. Boopy recalled that "the Balsan family had a very large chateau in the middle of Châteauroux. He [Louis Balsan] was very pro-American. The family had a lot of buildings around the chateau, and Louis rented them out to Americans. There were a lot of stories about what was going on there, and believe me, a lot was going on." One of these "stories" necessitated a discrete cover-up by Forman to save wealthy French friends from public embarrassment.

"It involved the wife of an Air Force officer. The wife of a French man had become a very good friend of mine and has remained so over the years. I visited her recently in Paris. She came to our house and asked Red if he would transfer the officer, who was a pilot, to another base, taking his wife with him and hopefully ending the affair. Red transferred the pilot out of Châteauroux. In the end, it made little difference because the French couple's marriage fell apart, ending in a divorce."

"Boopy had a great personality; I never remember her being unhappy or despondent," Donna said. "She had an upbeat attitude, and was able to cope with any situation, and she was as cute as can be. Her nickname 'Boopy' was a perfect fit. If you remember the cartoon character Betty Boop, that's who she looked like. She had those big, beautiful eyes and cute hair."

There was nothing conventional about the Formans. They met in Salt Lake City where Red was stationed, along with Boopy's first husband, Charlie Adams, a pilot under Red's command. "It was then that Red first called me Boopy. It stuck. Everybody calls me that. It's been a long time since I've answered to Betty. I never answer to Betty anymore: it's Boopy. I met Red when Charlie was off on a mission. Red sent him off again on another mission shortly after we met. And that was it." Red had adroitly tucked his rival away for the duration, but there was still a war to contend with. Shortly after, Red was sent to Burma. It was another five years before Red and Boopy were married.

By the time the Formans arrived at CHAD, Red's reputation had preceded him. He was among the pilots who flew the C-47s and C-46s over "The Hump," a perilous 530-mile flight over the Himalayas that supplied the lifeblood that kept Chiang Kai-Shek's army alive. In 1948, the Soviet Berlin Blockade began and Forman embarked on probably the most intense airlift planning he had ever undertaken. "Sterling Bettinger [Air Force Major] and I were flying the airlift commander, Major General William H. Tunner, into Berlin, and were unable to land since the runway had been temporarily closed because of an accident," Forman told his Memphis hometown newspaper, *Press-Scimitar,* on June 17, 1966. "When we finally landed, General Tunner kicked us off the plane, told us not to come back until we had devised a safer system that would work. We did."

The air lift orchestrated by Forman and Bettinger is universally credited with breaking the Soviet blockade and saving Berlin. For their effort, Forman and Bettinger were awarded the Legion of Merit.

In 1950, Forman led the airlift evacuation of 4,600 casualties from the Chosan Reservoir in Korea, a five-day operation that further embellished his hero status. Forman gave up his seat on the last C-47 evacuation flight to a wounded Marine. Joining the remaining Marines, Forman fought his way out of the Chinese trap, earning the Silver Star, a cluster to his Legion of Merit, and two clusters to his Distinguished Flying Cross. This was the legendary stuff that transforms a man into a hero.

The hard-drinking, fun-loving members of the Forman retinue did not live by the rules of the GIs who cavorted from one bar to the next in downtown Châteauroux. So what exactly was their job?

As John Riddle described, the 1602nd Air Transport Wing and the 322nd Military Air Transport Wing commanded by Forman were not part of the CHAD command structure, but were "tenants" who in effect rented space at the sprawling facility.

"The base was owned by the Material Air Command (MAC). So they had the responsibilities for being the 'housekeeper', for want of a better explanation," Riddle said. "As the housekeeper, they were responsible for performing all of the functions of the base. As tenants, we utilized office space, sleeping quarters, and things like that. The 322nd had nothing to do with that except for a small support squadron over at Déols. Our mission was to support MAC planes that were flying into Europe from the States. Basically, we were servicing the Department of Defense throughout Europe, the Middle East, and North Africa."

HIGH LEVEL HI-JINX

With a virtually independent status at CHAD, Forman was in the unique position to take advantage of the perks that came his way. One of them, a C-118 four-engine Liftmaster, would become an aerial mainstay for junkets. Riddle explained how this gift horse fit so well and could be comfortably saddled for multipurpose flights.

"It was about the time the Berlin Wall was going up [in 1961]. With all this going on, General Forman put out the word that he needed a plane. We were given the use of a C-118 for several years. We would get the plane and use it for about a year and a half until it was due to go back for maintenance. When the plane was sent back to McGuire [Air Base] for maintenance after three hundred hours of flight time, they would loan us another plane to take its place."

A tour of the chateau that would become the Forman home for six years revealed the muted splendor that greeted them in 1961. It was the former mansion of the Guillon family, who had operated one of the largest farm implement companies in Europe. A graceful staircase rose to the second floor from a beautifully designed terrazzo entrance emblazoned with a richly colored emblem that could have been the Guillon marque. It was a great place for parties, as was the isolated, former de Lesseps hunting lodge leased by the Hildebrands. When Donna and Hildy inherited the lease to the hunting lodge from Dick and Kay Wood, it had already won accolades.

"It was a great place for parties, a beautiful house," Boopy recalled. "There was one party that was so boisterous and bad that it still sticks out in my mind. The house had a very large dining room with a huge gold-framed mirror that covered one entire wall and ran from the floor to ceiling. The table was filled with hors d'oeuvres and all kinds of food. Somebody took some food from the table and threw it at the mirror and it stuck. With the group that we had you can imagine what happened next. Before long the entire mirror was covered with splattered food." Listening to Boopy, it was hard to get your thoughts around the fact that she was talking about high-ranking Air Force officers, including a general and lieutenant colonels, as well as high-ranking civilians lucky enough to be invited to the melee. "We all liked to party, party hard."

The hunting lodge satisfied a multitude of tastes. Donna Hildebrand, often described as a most innovative hostess, said, "We had a wine cellar with a couple of tables and the only light was from candles. We found out after a while that when a couple would go down there for some wine, they were taking a lot more time than it took to get wine. We discovered that the wine cellar had become a favorite trysting place, quite romantic." Kay Wood noted that romance did not begin and end in the lodge's basement. "There was also a stable and a garage that were popular. We also had a picnic area and a wooded area."

The floor-to-ceiling dining room mirror might have been the most massive piece of wall décor, but the expensive artwork in other rooms offered a bigger selection for creative hi-jinx. "I remember one party in particular," Riddle said. "Everyone who was there remembers it. Donna contacted the wives of all the officers who were invited and asked them to send her personal pictures. She then cut the faces out of the photos and pasted them over the faces in the portraits that hung all over the chateau. So when you arrived you would find your face pasted over the face of a guy in a hunting scene, for example. This was during the 1960s; everyone drank. So did we. The word 'wild' is relative. One man's wild is not another man's wild."

25.

Chateauneuf: Americans drove thirty miles to the base from this village.

A better than average kitchen "on the economy".

26.

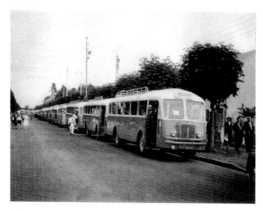

27.

28.

29.

30.

31.

32.

33.

34.

35.

36.

NOTES TO PHOTOGRAPHS 25–36

25. Touvent 410 Housing. Not all Americans lived in mansions. Teachers at the 410 housing complex put their students through their exercise paces. The apartment buildings built to French standards can be seen in the background.

26. Living on the Economy. As the number of American military and civilian personnel started arriving in droves, French farmers and urban landlords started converting properties to get their share of the American dollars.

27. It Started with a Broken Down Taxi. Abel d'Aquembronne recognized gold when he saw it. He parlayed a broken down taxi in need of constant repair into a transportation empire exemplified by this seemingly endless row of brand new buses that served most of County l'Indre. He could thank a sweetheart contract with the Americans transporting French workers to CHAD. Photo courtesy of Antoine Price.

28. Good Looks Could Pay Off. French workers at CHAD found there were many ways to work the system. Female workers at the base who were smart, good-looking and well-connected collected enviable perks. Here Lillianne Diez climbs aboard a C-47 for a weekend shopping and party trip to Paris, or was it Berlin. There were enough of them that Lillianne couldn't remember the destination of this trip. Photo courtesy of Lillianne Diez.

29. F-86 Sabre. Yes, there were NATO demands that had to be met at CHAD. The first big job to give hope to unemployed French aeronautic workers was the assembly of the F-86 Sabre fighter planes shipped in crates from Italy and to be parceled out to NATO air forces throughout Europe. USAF photo.

30. Cold War Realities. Cold War intrusions could not be avoided. Air Force surgeon George Banning was ordered from CHAD to lead a rescue mission to battle ravished Congo. Here a sweat soaked Banning aids one of the more than 100 refugees crammed aboard a C-130 at Stanleyville. USAF photo

31. Lunge, Parry, Thrust. Fencing commands entered the lexicon of the Golden Ghetto's sports monolith that welcomed even the most obscure games. Here Fencing Master, Phillipe Membre imported from Paris, instructs two neophyte swordsmen. The fencing team which included two airmen fresh off American farms, competed at some of the most exclusive fencing clubs in France.

32. Riding to the Hounds. Cries of "tally ho," seldom ever heard by middle class GIs back in their hometowns, resounded across the fields surrounding the Golden

Ghetto. Furious pursuit of the fox led the U.S. Dept. of Agriculture to warn riders away from French farms and pastures where hoof-and-mouth disease might be lurking.

33. CHAD Sports Summit. With two borrowed players the base basketball team came within one point of winning what was considered the European championship in 1955. NCAA Hall of Fame Coach Dean Smith, center front, was one of the two borrowed players. Ali Kahn, world renowned playboy and Hollywood sex goddess Rita Haworth were part of the picture. Photo courtesy of William R. Smith, coach.

34. De Gaulle Closes his "Hotel." This cartoon by Yardley was indicative of how the American media and public felt about Charles de Gaulle giving the boot to NATO forces in France in 1966, a joke. The escalating Vietnam conflict with its ever increasing casualties and coast-to-coast anti-war protests had already grabbed the American consciousness. Cartoon courtesy of U.S. Embassy Paris.

35. Only a Memory. Iconic images of huge cargo laden Air Force transport planes symbolized an economic bonanza that faded away with the base closure. The final base commander, brusque spoken fighter pilot, Col. Francis J. Pope, made no attempt to hide his anger at the base closure when he admonished local officials and businessmen, "We are leaving you the best and biggest airport in Europe. It has everything . . . Now, do something with it!" Official USAF photo.

36. Flame of Friendship (*La Flamme de L'Amitié*). Dallas artists Tomas Bustos, on the left, and David Newton, on the right, flank their creation at its dedication in 2010 in downtown Châteauroux. It was commissioned by an anonymous former CHAD high school student who wanted it to be an enduring tribute. (Author's photo)

Nothing was too childish. "Another time, I got hold of a mannequin and put it in the bathtub in the bathroom," Donna said. "No one was told about this so when the people came up to use the bathroom, the first thing they would see was this strange head peeking out of the bathtub. It startled a lot of the women." As Riddle recalled, one of the revelers could not let it go at that, because it just wasn't funny enough. "It was in the winter. The dummy started out the night fully clothed on the bed where it was eventually buried under the overcoats of the guests. Later, the dummy was stripped and placed in the bathtub," Riddle recalled. "Later, one of the guys came down wearing the dummy's clothes, explaining he put them on after spending time in the bathtub watching everybody going to the toilet. Of course, he didn't do that, but he got a big laugh."

The second anecdote is multifaceted, offering a mélange of pranks that sum up how far officer-class entitlement could be carried. It happened at a fashionable and expensive hotel in Vichy, which had become a favorite weekend golfing destination. Boopy described one incident without a hint of self-consciousness: "I was standing on the balcony of our room at this very nice hotel, and down on the street below me were Jay, my husband, Bob Goggin, and some others in our gang. They were trying to turn over a Volkswagen that was parked in front of the hotel.

"I don't know why they were doing it, they were all drunk. So in order to distract them before the gendarmes arrived, I began taking my clothes off on the balcony. I did a strip tease. My children can tell you about that because they were with us at the time. I can only guess what they were thinking when I stripped. Yes, I was very much naked. Luckily, nothing much happened, the gendarmes arrived. I knew someone would have called them. My husband and the others ran away before the gendarmes got there."

Susan Forman acknowledged that the balcony scene was hardly Romeo and Juliet. Despite being the youngest of the three Forman children, she was called "the little general" as a kid. She retains a vivid memory of her mother's balcony antics, also witnessed by her brother, John, and her sister, Joan. "We did see her performing the strip tease.

The three of us were in the next room and she was out on the balcony, and there she was stark naked," Susan said.

"I really don't know why she did it. All I know was that my dad was down in the street trying to get her to go inside or at least put something on."

The drunken zaniness continued into the night. "There was this poor unfortunate man in the hotel elevator," Kay Wood recalled. "I think he was one of the gendarmes. The elevator was one of those glass ones. Some of the men in our gang fixed it so he could not get out. I don't know how long he was in the elevator, but he did spend some time helplessly going up and down, while the guys just sat and watched him, drinking and laughing."

Kay Wood said that there was more than one slice of elevator hilarity that weekend. "I do recall a big ruckus over the hotel's elevator, and I do mean 'over' the elevator. Some of the fellows took control of the elevator and were climbing all over it. They opened the door at the top and climbed out the top of the elevator and began riding it up and down the exposed elevator shaft. Remember these were the old French elevators that were open, and you could see everything. They were made with this filigree wrought iron. It caused quite a ruckus, as you can well imagine."

Pretty Easy Duty

Kay Wood had been at CHAD since 1957, operating the base Service Club. She met Dick Wood while she was running a Service Club at the Air Force base in Rapid City and he was serving out his time in the Air Force. Wood completed his college degree, joined the Civil Service, and was assigned to the Air Force Special Services. He and Kay were reunited at CHAD, got married at Déols City Hall, and then moved into the de Lesseps Hunting Lodge before turning it over to Donna and Hildy Hildebrand. By that time, it had already earned an envious reputation as an ideal party palace. When Kay and Dick Wood handed over the keys to the Hildebrands, the de Lesseps property included two rental apartments that were always filled. Dick was the base recreational director at CHAD, a post that fit well into the Forman group.

Organizing trips, many trips, and all of them at Uncle Sam's expense, was Wood's responsibility.

"We had these wonderful conferences at great retreats in Europe. For instance, there was the R&R [rest and relaxation] complex at Garmisch, Germany," Kay said. "It's a very well-known ski resort area. We had taken over all of the resorts in the area for our conferences and retreats. That's when the State Department would send over celebrities such as Jack Nicklaus and Arnie Palmer and people like that, and we would set them up in the very best places. I realize that Air Force duty in France at that time was pretty easy duty, and R&R hardly seemed to be needed. I think it was just a leftover term from the past."

Members of the group loved to take photos. There were shots taken on the golf course, others at ski resorts—Austria and Switzerland were favorites—and there were before and after group photos with people wearing golf or skiing togs seated at tables with smiles on their faces and drinks in their hands. "Yes, golf was very popular, and in the wintertime we all liked to go on skiing excursions," Kay said, "but you would have to say that drinking was our most popular sport.

"I'll give you an example. It was on a skiing excursion to Switzerland. A bus was requisitioned for the trip by Red Forman. An enlisted man, I guess you could call him Forman's aide, set up a big bar in the rear of the bus, and it was in use constantly. There were base commanders from other commands in France making the trip with us. As you can see, we had no trouble having a good time."

The fun things the younger set did in and around the Golden Ghetto clearly defined who they were. Jean-Claude Prôt was taught baseball while living near 410 housing at Touvent, and after a game, the American kids drank Pepsi while Prôt and his French buddies collected mushrooms from the playing field for dinner. His wife, Nicole Prôt, formerly Nicole Neveu, rummaged the waste cans at Touvent, hoping to keep alive her dreams of a better life. She tore pages from American fashion magazines and stuck them on her wall and reclaimed high heels discarded by American women.

Red and Boopy's only son, John, did not inherit his father's flying gene and had no intention of being an Icarus who tempted fate with aerial stunts. He joined the Air Force in 1969 after being drafted, preferring a four-year enlistment to two years in the Army, possibly in Vietnam. He was a teenager attending CHAD High School when his father recruited him for party duty at the Forman chateau.

"I don't know whether I should tell you all about the parties at the chateau. I only had one job at the parties: bartender. My dad would have me tend bar for all of the people that would come in. I turned out to be a pretty good mixologist. I made everybody's drink except his."

When asked to describe a humorous incident during his high school years, John's answer was clearly indicative of the lifestyle he and his sisters enjoyed. "One of the funniest stories that I can remember is when I was with one of my best friends, another player on the JV basketball team," John recalled. "We were on a trip for a game in Paris, and the half-way point was Orleans. We all got out of the bus, all of the cheerleaders and all of the players, and I had this little camera. I had one of these little 'spy' cameras.

"While the guys were lined up outside, all of the girls went into the bathroom and I took a picture of them. When they found out, they were not too happy. My best friend chased me onto the bus, took the camera, and destroyed it. In retrospect, it was all very funny."

Susan Forman was only seven when the family moved to CHAD and barely a teenager when the base closed, but her parents made sure that she and her brother and sister were not excluded from the socializing.

"We were always introduced to the guests at the parties. At our house we had a powder room just off the main room. It was also a bathroom. Any time anyone would walk in there, we would all run upstairs, and if you flushed the upstairs toilet, the water would explode in the toilet downstairs and anyone would get soaking wet—I'm talking, of course about anyone, who was in there—and they would come out of the powder room and my dad would shout upstairs, 'John, Joan, Susan, get down here right now!' We would come down and everyone would end up laughing. That's what I loved about our parties."

Susan and John throughout their lives always referred to their mother as "Boopy." They never questioned whether it was appropriate or not.

Simply stated, it fit. Their mother was a very pretty, intelligent woman with a magnetic personality. From an early age, they recognized that their mother did not behave in lock-step with other officers' wives. John recalled an incident that summed up his mother's personality.

"I was with her on one of her little shopping trips to the commissary. There were continuous arrows painted on the floor up and down the aisles indicating which way the foot traffic should flow," John said. "My mother was going against the grain. A man—I believe he might have been the commissary manager—came up to her and said, 'Lady, don't you see those arrows?' My mother, without even breaking a smile said, 'Hell, I don't even see the Indians.' That little incident fascinated me. I was only fourteen years old, but even then I thought it was hilarious."

Susan said that when it came to discipline, Boopy and Red had a code of conduct that could not be trifled with. "I came home one day and walked in the door, and she was standing at the top of the stairs. When you walk into the house, that big staircase is right there. I was in the fourth grade. And there she was, my mom standing at the top of the stairs with her hands on her hips.

"She shouted down to me, 'I just went up to your room and found your Barbie bathtub with your cigarette butts in it.' Oh my God, here I was in fourth grade and I almost died at the bottom of the steps. When my mom caught me, I'm not even sure what happened at that instant. I'm not sure if I fainted or blacked out, but I did know that it wasn't funny."

With their father a general and their mother an iconoclastic, no-nonsense woman, the three Forman kids were young military royalty, with privileged insights and easy sympathy for those less fortunate. Theirs was a world that demanded deference, and when it was lacking there were consequences, as they witnessed when returning home from the La Martinerie golf course.

"That's when my father was pulled over by an air policeman for speeding. The three of us were sitting in the back when my father called this airman over to the car and asked, 'Do you know who I am?' We were all sitting in the back with our eyes as big as saucers, and we all felt so sorry for that young airman. It was obvious he didn't know who he had pulled over and he said, 'Well, Sir, you were speeding.' Our father said, 'I'm going home to drop my family off and I will be back to talk to you,' and we knew that this poor guy was in trouble," Susan recalled.

John's sisters were dropped off at the chateau, but he remained with his father, returning to where the young air policeman was waiting as ordered beside his patrol car. "My father pulled rank on him. He told him to get back into his car and follow him to the Provost Marshall. We walked into the office and my dad said to him, 'Son, do you see that picture up there on the wall, that's me. No matter how fast I'm going, even if I'm driving down the middle of the flight line, you don't stop this car.' This guy had no idea who my father was, but you could just see that he was petrified. I found out that my father went back to the Provost Marshall and had the air policeman promoted the next day. That's the kind of man he was."

Then there were all those fun trips enjoyed by the Forman clique aboard the C-54 Skymaster the general had requisitioned in order to inspect the various NATO air bases serviced by the 322nd Air Division. There were no problems when the official tour of inspection ended, and the Skymaster just happened to touch down at an enticing and exotic city along the way. Beirut was a big favorite. "We would fly to all of the bases under Red's command in the Mideast. I don't know whether these trips were authorized or not, but we would fly to Ankara, Turkey, and on the return we flew into Beirut," Boopy said. "Those were the days before the civil war in Lebanon. Beirut was a wonderful city, a divine place. We had many wonderful times there."

"I was dating the secretary of our ambassador in Lebanon [Armin Meyer] and was driving his car to the Phoenician Hotel in Beirut, a really great spot to meet her. It was where all of the stewardesses from the major airlines stayed," said Jay Parsons.

"The Phoenician Hotel—I remember it quite well," Boopy recalled, "A wonderful swimming pool and great bar."

In what became a vacation destination for two summers, the Forman entourage either flew aboard Forman's personal C-54 or drove to

Spain's Costa de Brava not far from Barcelona. Originally, boredom was the motivating factor.

"Every summer, the kids would get tired of hanging around in Châteauroux all the time. After a while, visiting France's cathedrals and castles and places like that got kind of old," Boopy said. "We started going to Spain, and every summer we would take a house for the entire summer. Red would go back and forth to the base, of course, and Jay would be with him on those flights."

John Forman recalled that a favorite activity was driving golf balls down over the village of Blanes, bouncing the pellets off the rooftops to the extent that the mayor drove up to the Forman house to complain.

"There were a lot of parties there, a lot of things going on," Susan said. "I remember Jay Parsons one time when all of us kids accidentally caught him in the back seat of a car with a woman. Yeah, they were doing what you would expect them to be doing, but I was so young I really didn't understand. But my brother and sister, who were older, giggled a lot."

Séances and Pink Ladies

Social decorum was hardly a consideration, nor was there much thought given to the outrage their lighthearted pranks might provoke. When Bob Goggin took on the role of a mystic hypnotist in a restaurant, it made little difference how much heartburn his fake séance might cause. Although Goggin was one of the youngest members of the Forman ensemble, he was acknowledged to be one of the funniest. "We would go into bistros in downtown Châteauroux and have bottles and bottles of wine on the table," Boopy said. "Bob Goggin would then hold a séance. He would have us put our fingers under the table and then start the séance by having us lift the table from the floor like it was floating. And the poor French people in the bistro, some of them eating dinner, would just stare at us shaking their heads. We were quite terrible."

It is doubtful if any of the bistros and restaurants frequented by these high ranking revelers kept handy the telephone numbers for the air police and local gendarmes. Conversely, CHAD enlisted men learned very quickly that they were on a fifteen minute leash; if a hostess bar or saloon owner even suspected that there might be trouble, arguments, and fistfights, it would have to be settled quickly before the air police and possibly some gendarmes arrived.

At CHAD, the Officers' Club offered the Forman entourage, other officers, NATO members, and civilian employees a watering hole designed to satisfy all but the most insatiable. Entire families enjoyed the club's dinners, dances, musicals like *Guys and Dolls*, anniversaries, and costume parties, while the more permissive parents would allow their kids a pull or two at the slot machines. "We had striptease every two weeks on Friday. The women were not allowed to come, only the officers. Professional stripteasers would come from Paris to do the show," said Francis Bayard, a longtime bartender at the Officers' Club. "This often distracted my French employees from their chores, coming out of the kitchen to take a peek!

"The U.S. officers were different with their families, less strict with their wives and children," said Bayard, who witnessed "a lot of jealousy between U.S. officers and French officers, especially on the French side. The living conditions and the salary of the French officers were far different. This created tensions. At the club, we served Manhattans, Martinis, Pink Ladies, and all the different whiskies, scotches, bourbons, and cocktails that the French ignored the existence of—the only place you could find these drinks besides the base was in the plush restaurants in Paris."

Drinking was the liquid core of the high octane, combustible environment that dominated much of the club's activities. In this boozy arena, even a serious drinking problem could easily be reduced to a joke. As described by Jay Parsons, the heavy drinking of Colonel Joseph Nagle, a chief flight surgeon and commander of the 7373rd AFH, was an example of how inebreation could be laughingly accepted. "It was all done in fun, sort of a joke. Joe loved to drink, sometimes too much. So we attached an airplane safety belt up to his favorite stool so he wouldn't fall off onto the floor."

"It was very sad to see, very sad," said Boopy of Nagle's drinking. There was no evidence that Nagle performed surgery, only that he had the title of hospital commander.

When the club concoction included a dance band, a good-looking French woman, and a snoot full of booze, it proved too much for young dentist Captain Josh Getzoff to handle. "Colonel Gibson, I can't remember his first name but everybody naturally called him 'Hoot,' after the old-time cowboy movie star, lined up a young French lady for me, a blind date. Usually I don't drink very much, but I got hammered that night. I ended up doing some pretty bizarre things with her on the dance floor. She was drunk, too. I could hear comments from the people watching us: 'How disgusting for a doctor to behave like that,'" Getzoff said.

"Finally, the officer in charge of the air police, Lieutenant Charlie Bachelor, came over to us and said we had to leave. He was a Southern boy, and when he came over he said, 'Well Josh, I think you'd better get your ass out of here, you're making a fool of yourself. Man, you are just too drunk.' And just like that I got my cool ass out of there."

It seemed that no matter where you turned at the Officers' Club, mind-numbing elixer could be found. "We had an awful lot of special banquets at the club, and typically the dessert was Baked Alaska," Club manager Whalen said. "The Baked Alaska was pretty big. Sometimes, with a big party there would be two of them.

"We would pour hundred proof vodka over the Baked Alaska, turn out all the lights, ignite it, and then wheel it in on a cart. It was very dramatic," he added. "One night we did everything right and tried to ignite the dessert, but nothing happened. So I said 'God damn it, let's roll it out anyway.' When it came out it was just a big soggy mess. Nobody was happy. I looked into it and discovered that the cart with the Baked Alaska was next to the dishwashing machines, and the French dishwashers had guzzled all of the vodka, two bottles of it, and replaced it with water. So we had a waterlogged cake that nobody could eat, and in the back room, four drunk Frenchmen who could hardly stand up."

For Red, Boopy, and their acolytes, the Officers' Club provided yet another avenue for what appeared to be an around the clock need for escape. This did not deter them from their Cold War mission, of which Forman never lost sight.

Humanitarian aid was rushed from CHAD to Yugoslavia, where an earthquake destroyed 75–80 percent of the Macedonian city of Skopje, killing 1,070 people and leaving 100,000 to 200,000 people homeless. Among the supplies aboard the planes dispatched by Forman was a complete field hospital with drugs that quelled an epidemic.

When a bloody civil war broke out during the early 1960s in the Belgian Congo, the United Nations poured in thousands of troops from Ireland, Sweden, Ethiopia, and Indian Gurkas. All the Belgians fled the country. The U.S. Air Force was responsible for conducting a massive airlift required to sustain the U.N. forces and feed the beleaguered population. Military supplies, food, and medicine flew into Albertville, Elizabethville, and Leopoldville around the clock. It was no great surprise that Red Forman, just as he did in Korea and during the Berlin Airlift, was selected to oversee this dangerous operation.

Forman retired on March 1, 1966, just a few days after returning to the States from Châteauroux. He was killed on June 19, 1971 when returning to his Philadelphia area home from the U.S. Open Golf Championship at Merion Golf Club in Ardmore, Pennsylvania. Boopy, who had remained home to tend to her sick daughter, explained: "His car was hit broadside. The driver of the other car apparently had a heart attack and his car jumped a median. Red was killed instantly."

1963 – COLD WAR POTPOURRI – 1963

- On May 7th, Asst. Defense Secretary Sylvester predicted, "The corner . . . has been turned toward victory in Vietnam." And General Harkins, U.S. Vietnam Commander agreed that ". . . the end of the war is in sight."

- On January 11th, the nation's first Disco club, The Whisky a Go-Go, opened on Los Angeles's Sunset Strip, kicking off an international music and dance craze that would last for more than a decade.

- On August 28th in Washington D.C. the Reverend Martin Luther King, Jr. delivered his "I Have a Dream" speech from the steps of the Lincoln Memorial to more than 200,000 people crowding the mall below.

- From June 15–16th, Astronaut Gordon Cooper orbited Earth twenty-two times in the last Mercury mission, and a month later the Soviet Union's Valentina Tereshkova orbited earth for three days, becoming the first woman in space.

- On January 29th, Sam (Slinging Sammy) Baugh and Bronco Nagurski headed the list of the first seventeen former players inducted into Pro Football's Hall of Fame.

- On June 26th, more than a million West Berliners cheered President John F. Kennedy when he told them "Ich bin ein Berliner" (I am a Berliner), extolling his listeners for defining the meaning of freedom.

CHAPTER 13

MOURNING, MINDSZENTY, AND MICHELIN

They carried in a casket draped with an American flag. Carrying the cas-
ket were former members of the French underground, the Maquis. It was
a tremendously moving experience.

—Air Force Captain John Riddle

It was November 23, 1963, and Lieutenant Robert Goggin and a fellow JAG officer from CHAD were searching for a way to return to their home base. After scratching around, they considered themselves lucky to have uncovered an Air Force sedan that was somehow overlooked in the motor pool at the Wiesbaden Air Force Base in Germany. It was one of the few vehicles at the huge Air Force command post that had not been pressed into emergency service when all NATO bases in Europe were placed on a twenty-four-hour alert after the assassination of President John F. Kennedy in Dallas the day before.

The two junior JAG officers had been scheduled to fly home after completing a legal matter, but getting a military aircraft of any size was out of the question. Instead, they faced an auto trip of more than eight hundred kilometers through regions of France with which they were unfamiliar. After crossing the French border on route to Metz, they saw for the first time a phenomenon they never would have expected. Driving through the small farming village of Freyming-Merlebach, they found that the buildings, from the smallest farmhouse to the village *La Marie*, displayed flagpoles draped in black bunting and the French tricolor. It would be hard for Goggin to imagine that the scene had been choreographed, but as he and his companion passed through cities and towns, their original sense of wonderment experienced in Freyming-Merlebach turned to awe. It was as though a master director had placed his handiwork on display. Through Boucheporn, Metz, Nancy, the industrial town of Ste. Dizier, Poitiers, and on and on, it was all the same, a nation paying tribute to a fallen friend.

"It was chaotic, no one knew for sure what was going to happen," Goggin recalled. "But something uplifting did happen on our drive back to Châteauroux. In all these little French villages we passed through, they had the French tricolor flying with black bows tied to them. This was at almost every house we passed along the way. It was very touching, and showed us just how much affection the French had for the President. It was an amazing thing to see."

At the quadrangle at La Martinerie, both the American and French flags were lowered to half-mast immediately after news of JFK's assassination reached the base. Both flags were to fly at half-mast for thirty days, in keeping with American tradition after the death of an important political or national person, and as a tribute by the French. Traditionally, the French tricolor would fly at half-mast only until a state funeral, but at the request of Lt. Col. René Bordes, the French chief of liaison at CHAD, the rule was suspended so that both the American and French flags could fly side by side at half-mast for thirty days.

When Captain John Riddle and his wife accompanied his boss, General Red Forman and his wife, Boopy, up the main aisle of a packed Ste. Andre's Church in downtown Châteauroux for a commemorative tribute to JFK, they had just pushed their way through hundreds of French mourners who had found it impossible to get inside and crowded the paved and cobblestoned *place* outside. As they neared the sanctuary, Riddle was astonished to see that "big overstuffed chairs had been placed for us. The crowd was overflowing, filling the church and the area outside. Loudspeakers were set up so that everyone could hear the service.

"It was not long after the Forman party was seated that former Resistance fighters solemnly carried an American flag-draped casket up the aisle to place it in front of the sanctuary. Behind the casket, these former *Maquis* marched in carrying pieces of tattered cloth that were apparently their standards during WWII," Riddle said. "It was a tremendously moving experience. It demonstrated how deeply caring the French people were for us and the President that we had just lost."

CHAD, like every other NATO base in Europe, was placed on virtual wartime footing. It did not take long for shock and disbelief to evolve into sorrow and spiritual trauma. Over the decades, it has become a cliché to ask a person to remember the instant they learned of JFK's assassination. It was no different at CHAD.

"I do remember when John F. Kennedy was killed," CHAD High School student Annette Gagné recalled. "We were in the movie theater on the base when the announcement was made and it traumatized everybody. People were afraid it would be the start of another world war. It was unreal. Some of the guys at the theater began to cry. My boyfriend, Bill Thomas, and I started crying."

Pat Thacker, fourteen-year-old CHAD High School student, and daughter of Captain Jack Thacker and Mary Thacker, was watching *The List of the Adrian Messenger* at a base movie theater when, "the film stopped and the lights went up. A young airman spoke up from the back of the theater, 'The President of the United States and the Governor of Texas have been shot.' I was stunned, my friends and I were asking if it could be real. When the movie was over, we went outside to wait for the next bus to take us to our home in Brassioux. After I got home, I rushed in to find my mother and younger brother listening to the radio. We were glued to the radio that evening."

For Peter Nyberg and other CHAD High School students, the assassination elicited more than shock and disbelief—it also represented the unthinkable. "It wasn't just that a shooting had occurred, it was that our President had been killed. Things like this just don't happen. It was unthinkable."

For the CHAD officer class, the assassination was a call to action. If orders had to be given, those in command had to be in place in order to issue them. "My boss, Lieutenant Col. William Spizer, and some other officers were giving a dinner party at a very nice restaurant outside of Châteauroux," Donna Hildebrand said. "There were probably about forty or fifty people at the party, including the base commander. An officer was sent out from the base to tell us that President Kennedy had been assassinated. At first, we all were wondering how this could be true, and if it wasn't somebody pulling a stunt on us. We were finally convinced and, as you can imagine, the party broke up immediately and everyone headed back to the base as fast as they could in order to do their jobs."

The President's assassination created a problem for Marty Whalen. As manager of the Officers' Club, he knew that the restaurant and dining room must continue operating, but there was also the bar to be considered. The bar was a popular watering hole for not only CHAD officers and their wives, but for NATO personnel, nurses, teachers, and Civil Service employees as well. Whalen rationalized that a stiff drink or two could ease the pain caused by JFK's death and decided, not without misgivings, to keep the drinks flowing. As a result, his decision unwittingly contributed to the only ugly incident that anyone can remember surfacing in the aftermath of the Dallas assassination. John Riddle recalls what happened:

"There was a transient crew that had just flown in from the States. They were sitting at the bar, which was pretty empty at that time," Riddle said. "One of the pilots was from Texas and he made a toast to the new President from Texas [LBJ]. Also at the bar was the wife of a government Civil Service employee. She told the pilot that his remark was very inappropriate, very rude. Captain 'Rip' Fox, a pilot under General Forman's command, was also at the bar. He got into it with the Texan, throwing a punch and knocking him down," Riddle said. He couldn't remember Fox's first name, only that everyone called him "Rip."

"General Forman had known him back in the days when they were stationed at Evreux-Fauville. General Forman found out about it the next day and called the transient crew's commanding officer back in the States," Riddle said. "He said he would personally handle the matter, holding it to the minimum so that it did not get out of hand. It ended up with a letter of reprimand for Rip, which he was not very happy about. But he came to realize it was the best thing, because this was not permanent and was expunged from his record when he was transferred."

French radio and television gave the assassination round-the-clock coverage and the regional daily newspaper, *La Nouvelle République*, emblazoned its front page with the largest headline its typesetters could muster: "*KENNEDY ASSASSINÉ.*" The voices of the announcers for Armed Forces Radio were tremulous and the base newspaper, *Sabre,* was filled with personal recollections of the moment that the impossible became reality, that JFK had been shot to death by Lee Harvey Oswald. One officer observed that "a kind of fear seemed to grip everyone there; a sort of quiet panic."

At least one Cold War barrier disappeared in the wake of the tragedy, if only momentarily. Soviet Premier Nikita Khrushchev cut short a tour of the Ukraine and returned to Moscow where he signed a book of condolences at the U.S. ambassador's residence, and declared that Kennedy's death was "a heavy blow to all people who hold dear the cause of peace and Soviet-American cooperation." Charles de Gaulle, who turned seventy-three the day JFK was killed, and who escaped several assassination attempts himself, said, "President Kennedy died like a soldier, under fire, for his duty, and in the service of his country. In the name of the French people, a friend always of the American people, I salute this great example and this great memory."

Hungarian Dissonance

The Kennedy assassination marked the second time in just over seven years that CHAD and every other NATO military base in Europe was placed on twenty-four-hour alert. Bloody events in Hungary caused fear to spread from one Western European country to another that the Cold War was about to turn hot. On October 23, 1956 an estimated 20,000 protestors, mostly students, took to the streets of Budapest vowing, "we will no longer remain slaves" to the Communist regime of Ernö Gerö. By six p.m., 200,000 protestors were in the streets, and two days later, Gerö was forced to resign as Soviet tanks and troops moved into the city. Over 2,500 Hungarians and 700 Soviet troops were killed before the uprising was put down on November 10, 1956.

There were no condolences from Khrushchev, who had seized power three years earlier and viewed the Hungarian uprising as a threat to Soviet hegemony in Eastern Europe. Charles de Gaulle was out of power, writing his *Mémoires de Guerre* at his home at Colombey, and Jack Kennedy was into the fourth year of his first term in the Senate. At

CHAD there was uncertainty, as daily routines were disrupted, giving rise to both fear and annoyance. Much has been written about the massive NATO retaliatory power that stood at the ready during the three weeks that the Hungarian crisis played itself out.

It would be hard to find a more cogent, grassroots view of the crisis than the perspective supplied by Helen Pattillo in letters home to her parents, Walter and Zylphia Goodart, in Salt Lake City. An overriding tone of exasperation and domesticity disrupted, along with fear and anxiety, is evident in her first mention on November 7, 1956 of the Hungarian crisis:

> Dee Banning's eight-month-old baby is sick. Her husband, George, is the surgeon who was sent to Budapest with U.S. Ambassador [Edward T.] Wailes, along with medicines, medical supplies, first aid equipment, etc. Now that he's in Budapest, there's no way to hear from him and naturally, Dee's worried. Those Russians are rough. Goodness knows what's going to happen. With all that's going on in Hungary, and the Suez Crisis, the general [Pearl Harvey Robey] here started asking about our evacuation plan, and apparently there isn't one.
>
> It was supposed to have been set up two years ago, but fell through the cracks. American military forces in Germany have such plans and test them every few months, but people here have now started scurrying around and will conduct their first test next Friday morning. Unfortunately, our apartment building is one of those selected to participate. The boys are unhappy that they can't stay out of school and participate. I must drive to the base, get to the assembly point at nine, and then to the theater for a briefing. (You'd think they'd brief us first, wouldn't you?)

Helen went on to describe how the CHAD Director of Plans and Programs stopped by the apartment and "explained the whole thing to me." There was no mistaking that a sense of incredulity prevailed when she heard what would be required before the family could join an evacuation convoy of 1,500 cars, along with gas trucks.

"I can't remember shopping in Châteauroux for anything except toiletries," Helen said, laughing. So when the list of required evacuation essentials was handed to her, her shopping universe expanded enormously. It included:

"Three blankets per person, warm clothing, food for three days, knives, forks, spoons, plates, cooking pans, flashlights, batteries, money, changes of clothes in one suitcase per person, itemized list of household goods & personal effects, automobile spare parts, Thermos jug, passports, control cards (ten per person, thirty for the three of us), First Aid Kit, road maps, etc. I'll do as told. I'm glad we're coming home soon, and won't have to go through this exercise too often, much less the real thing."

One week later, on November 14, it became apparent that dependents at the base felt they were being kept in the dark, that military authorities were rationing out information, and they were receiving only second-hand news over the radio. However, nobody was losing sight of the fact that, should the shooting begin, they had to get out of CHAD, and possibly France, as fast as they could.

Helen wrote, "As for the war situation as seen from here, we learn only what Armed Forces Radio tells us, and are too close to be objective. We dug out the radio to listen to BBC at night—the only time we can get it. All leaves are canceled. They have attempted to practice the emergency procedure by which civilians are supposed to get out of here, and I fail to see that it would work in any respect. Roads are always mobbed in an emergency in this country; there are so many farm vehicles, bicycles, etc., that the roads would be impassible—especially for large, heavy American cars.

"Things seem a little calmer than they were a few days ago. We keep our things where we can reach them in a hurry. There are rumors they are working on a plan to fly us out, if possible. That would make better sense. We are holding onto as much 'green' [American money] as we can, and buying travelers checks with the rest, in hopes we can pay our way onto a ship or airline if we decide to leave. If the situation in Hungary settles down, we may fly over to England for a few days," Helen advised her parents.

A letter that Helen's close friend, Dee Banning, received from her husband, George, written upon his arrival at the American Embassy in

Budapest on November 7, added to their overriding anxiety. As far as the Soviet Union was concerned, what was happening in Budapest was anything but a training or practice exercise. As the death toll among Hungarian dissidents mounted, there was little doubt that Communist troops understood quite clearly why they were sent to Hungary. As Helen described to her parents, Banning, who would later serve as deputy commander of the 7373rd AFH upon his return to CHAD, found himself in the middle of international intrigue.

"George Banning called the hospital here from Budapest Tuesday, which was the first word he has been able to get out of there. Since then, Dee has received a letter written November 7, right after he reached Budapest. He will be there another three weeks. He was in the last car to get through and, although they were not physically injured, they were stopped frequently and guns poked through the windows at them. He set up a hospital in the basement of the Embassy. There are sixty-five people living there. He's in charge of food rationing, and had enough for two weeks from November 7. The Egyptian Embassy was completely demolished in the fighting, the city is a mess, but no one has bothered the U.S. Embassy."

Among the refugees sequestered in the Embassy's basement was Catholic Cardinal József Mindszenty. The prelate was released from a Communist prison on October 30, 1956, seven days after the street protests erupted in Budapest. His release ended almost seven years of imprisonment that began after his torture-induced statements at a show trial caused him to be found guilty of crimes against the Communist state. He arrived in the city on November 2 and, after making a radio broadcast praising the Hungarian insurgents, was forced to seek asylum at the U.S. Embassy. He would remain in the Embassy for fifteen years, despite repeated demands by Communist authorities that he be handed over to them. During that time, he became a Cold War symbol of anti-Communism, and a liability to both the East and the West whose diplomats never quite agreed on what to do with him. Finally, on September 28, 1971, he was allowed to leave the country and went to live in Vienna, Austria. George Banning never forgot the cardinal, and considered their brief association at the American Embassy to be one of his most memorable military experiences.

"My father had many anecdotes describing Mindszenty," said Debbie Martin, who was only two at the time of the Hungarian uprising. "My dad was quite impressed by the man, and spoke often of his dignity and warmth. He told us the Cardinal was obviously the most prominent refugee who had sought protection in the American Embassy. I don't recall if my dad personally treated the Cardinal, but my dad had been sent there as a doctor, and he came under the direct supervision of the American ambassador. I guess you could say, a doctor in residence."

The medical assistance that Debbie's father gave to the refugees at the American Embassy in Budapest proved to be a tepid practice run that could hardly compare to what Lt. Col. Banning and his small medical team from CHAD faced when they flew into Stanleyville during the bloody civil war that engulfed the Congo in 1964 and 1965. It was another case of the Cold War turning hot and bloody. Primitive Simba tribesmen, followers of the late Congolese President, Patrice Lumumba, had taken over much of the Congo and were freely roaming through the countryside, slaughtering real or perceived colonialist sympathizers. Lumumba was believed to be a Communist sympathizer by the U.S. government, and as such, had to be eliminated. Overthrown in September 1960, Lumumba was killed by firing squad in January 1961. One of the Simbas' massacres had occurred in Stanleyville. American and European survivors straggled on foot onto the tarmac of the Stanleyville airfield, where Banning and his small medical team waited beside a C-130 evacuation plane.

Banning's team used morphine to quiet the wounded while IV bags were set up along the length of the C-130's interior. One photo showed a blood soaked Banning bending down to assist a wounded refugee on the deck of the plane. Banning's plane was under heavy fire from Simba tribesmen when it was ordered to take off. A hundred wounded refugees were flown out to Leopoldville. Despite efforts to keep them alive, three of the refugees died.

Banning was apparently the kind of guy who was hard to forget, and for all the right reasons. Air Force Major John Robinson, whose public affairs team accompanied Banning from France, was aboard the evacuation flight from Stanleyville. Robinson, whose crew put down their cameras to assist Banning's small, overworked staff aboard the flight to safety, said, "George was very cool under pressure, nothing rattled him. And he had more charm than a barrel of monkeys."

"George was a charming guy, a lot of fun," said James Pattillo, "and quite handsome, very trim, had a great smile. I had already left CHAD when he became deputy commander of the hospital, but knowing him, I am certain he did a great job."

ABSTRACT TRANSFORMED

For the pilots and other officers attached to Forman's entourage at CHAD, the former hunting lodge of Ferdinand de Lesseps and the general's party-filled, mansion home were fast becoming no more than fond memories. During interviews, they seldom mentioned the Cold War. When the question of Vietnam arose, it was almost always addressed in the abstract. But abstract assumptions can soon become a reality if supported by years of bloodshed.

One of Forman's two aides at CHAD, Captain John Riddle, once rationalized the Forman clique's often irresponsible behavior by saying "one man's wild is not necessarily another man's wild." Forman's second aide, Captain Jay Parsons, was amazed at the easy availability of good-looking women. Ordered to Vietnam, Riddle found himself performing the job for which he was trained—flying a plane. Baked Alaskas were no longer the concern of Captain Marty Whalen, who found himself doing what he loved best while in Vietnam—sitting in the pilot seat of a troop and cargo carrier. Major Donna Hildebrand, the playful party organizer who once oversaw a wine cellar trysting place, served more than a year as personnel executive officer at 7th Air Force headquarters in Saigon.

Donna's husband, Lieutenant Col. Floyd "Hildy" Hildebrand, had retired from the Air Force before taking to the skies in Southeast Asia. He was not at the controls of a mammoth transport plane, but was flying a comparatively antique C-47 "Gooney Bird," courtesy of the CIA. The plane's identification markings were simple and direct, "Air

America." The two words symbolized for anti-war activists everything that was covert and possibly illegal about U.S. involvement in Southeast Asia, especially Laos. "He was out of work and looking for a job. When he signed on with Air America, it was a natural thing to do," Donna said.

From 1959 to 1974 the CIA's Gooney Birds performed air-drop missions and supplied logistical air support in Laos. Hildy thought it was absurd that people thought covert trickery was involved: "They called it a clandestine operation, but hell, everyone knew what it was. Anyway you look at it, I was a mercenary." Hildy joined a troupe of out-of-work U-2 pilots looking for a job. Most had flown out of a remote Beale Air Force Base in north central California, and among them was Rosey Rosenfield, a friend of Donna's, who described the camaraderie and bonding that the dangerous U-2 missions had created: "I was looking for a job and I was attracted to Air America," Rosenfield said. "When I got to Washington and sat down, and then later met with the Air America people, I found out that I knew just about all of them. It was like old home week."

John Riddle found himself at the controls of an unglamorous Air Force troop-and-cargo-carrying workhorse, the twin-engine C-123. For six of the thirteen months he spent in Vietnam in 1967–68, Riddle flew with the 309 Air Commando Squadron. The C-123 was well suited for the Air Commando missions Riddle flew. It was simple and rugged, could take off and land on rough airstrips, sometimes no longer than two thousand feet, and almost always came under heavy enemy fire.

One of his crewmembers for awhile was President Johnson's son-in-law, Patrick John Nugent, who was married to LBJ's younger daughter, Luci. Donna Hildebrand, who remained a friend of Riddle's long after their CHAD days, said, "John was a very smart man. When Luci's husband arrived at Tan Son Nhut, other pilots shied away, thinking having him as a crewmember would have the President looking over their shoulders, but not John. He told me, 'If he comes under fire at any time, he'll get a medal, and if I'm the pilot, well, who knows?'"

"Nugent was a capable crew member, a good loadmaster," Riddle said. "My C-123 was not the only plane he crewed. I remember him as a friendly guy who did his job well."

Riddle's missions during the fifty-day battle for control of the northwest highlands' Khe Sanh airstrip in early 1968 were typical. "We were in and out of Khe Sanh almost every day that the Marines held out at their fire base," Riddle said. "It was right up to the day the Communist siege ended and the Marines were relieved." Another beleaguered outpost was the small, central highland airstrip at Ban Me Thout. Riddle was awarded a Silver Star for making repeated nighttime landings under heavy enemy fire at the unlit airstrip to deliver troops and cargo. Besides the Silver Star, Riddle was awarded two Distinguished Flying Crosses (DFC), a Bronze Star, and five other citations for meritorious service. Riddle transferred from the active Air Force to the active Air National Guard, and retired in 1990 as a major general.

At CHAD, Red Forman's group had no trouble standing up and being accounted for, whether it was at a food-throwing, childish escapade at de Lesseps' hunting lodge, hitting golf balls from a Spanish hilltop into a seaside village below, or trapping a gendarme in the elevator of a fancy Vichy hotel and watching him ride helplessly from floor to floor while they toasted him with drinks. Fun and games behind them, they had no trouble standing up and being counted in Southeast Asia.

A Swimming Pool and Ambushes

Thousands of enlisted airmen passed through CHAD, with perhaps 90 percent of them describing the base as possibly their best Air Force duty station. Then there were the military brats, Air Force dependents who bounced along with their parents from one posting to another. Life was good. It was only because of exceptions like the faraway Korean War, and the closer-to-home Hungarian uprising that denizens of the Golden Ghetto were reminded that the Cold War could become hot and deadly.

Finally, it was the Vietnam War that lifted the Cold War out of the abstract for the CHAD "kids of summer," who only a few years after high school would be facing the jungle and Mekong Delta realities of Vietnam. It is impossible to say how many CHAD students found themselves in Vietnam. It is also impossible to construct a profile of a typical CHAD High School student who served in Southeast Asia. So I chose three of them, Mike Gagné, Peter Nyberg and Frank Nollette, to be the student standard bearers. All of them saw action, two were wounded, two of them received Purple Hearts.

Mike Gagné carried with him dual French and American citizenship when his plane arrived at Tan Son Nhut Air Base in Saigon in 1970. His first Vietnam assignment as an air cargo specialist was cut short when he volunteered for flight duty without knowing what he was getting into. There was an insatiable need for French interpreters as American military involvement spread clandestinely into neutral Cambodia and Laos. Only one day after accepting flight duty in Saigon, Gagné was strapping himself into the backseat of a twin engine OV-10 Bronco at the Rustic FAC Group's airfield in Bien Hoa. The versatile Bronco's primary missions in Vietnam were as part of counter insurgency combat, and as a spotter.

During his five months—June to October—with Rustic in 1970, he flew forty-six missions into Cambodia, at the same time President Richard M. Nixon and Secretary of State Henry Kissinger were insisting that none of this was happening, that such action was illegal, and therefore out of the question. "We all knew we weren't supposed to be there, but orders are orders," Gagné said. Gagné, like everyone else assigned to the Rustic Group, was told to "keep my mouth shut."

To get from Bien Hoa to Cambodia, the low flying Broncos passed directly over one of Michelin's sprawling rubber plantations, placing the surreal nature of the conflict into dream-like focus. "All around the plantation were big bomb craters, so you knew there had been plenty of action and a lot of people had probably died," Gagné recalled.

When Gagné's plane flew at an altitude high enough, he could see evidence of carnage up to, but never including, the plantation. Not far from one crater there was a big, untouched, French-looking chateau complete with large swimming pool. Flight after flight, Gagné saw Caucasian men and bikini-clad women cavorting around the pool. "They waved every time we flew over. They were used to us, we had flown over so many times," Gagné said. "They knew they had nothing to fear from us. Everybody at Rustic knew the plantation was protected and that a deal had been made. It wasn't fair, wasn't fair at all.

"It really bothered me that GIs were being ambushed and killed on the plantations at the same time Michelin's people, probably supervisors and their families, were having a good time around the swimming pool," Gagné said. "I never saw the fighting, but everyone knew it was going on. The plantations were big staging areas for the VC and the NVA. People at Rustic were throwing it around that Michelin got six hundred dollars for every tree that was damaged."

Despite his forty-six secret combat missions, often through withering ground fire, Gagné ended his tour of duty in Vietnam without a scratch.

During Peter Nyberg's childhood as an Air Force brat and later as a young adult, he had no misgivings about his family's way of life. But he did have doubt about whether the concept of military life, with its harsh demands for discipline and respect for authority, could ever be appreciated by those who had never experienced it. The son of a career Air Force non-commissioned officer, Nyberg returned to the States in 1964 and attended Arlington State College for two years, earning an associate degree. It was during those two years at Arlington College that Nyberg discovered for the first time that his reality was not necessarily shared by his peers and instructors.

"For example, when you discussed your experiences and how you interacted with people in France, Germany, and other places, they were somewhat lost," Nyberg said. "You really couldn't talk at length or in depth, because they really did not know what you were talking about. Those were the college classmates, and hands-on instructors, and professors. When you tried to discuss things with people, they were somewhat cautious, confused or distant. It was because you were discussing things of which they had no experience and no knowledge."

He joined the Army in 1966 and received a second lieutenant's commission in 1967 after attending OCS at Ft. Hood, the sprawling Army base in southwest Texas. Nyberg was shipped to Vietnam in 1968, taking over as a reconnaissance platoon leader with the Second Battalion of the Second Mechanized Infantry Regiment, First Infantry Division. He was immediately thrown into action amid the seemingly endless rows of rubber trees on the vast Michelin plantations about thirty miles northwest of Saigon.

When Nyberg's armored personnel carrier rumbled into the plantation on September 27, 1968, its mission as always was simple—to ferret out and kill the Viet Cong who had honeycombed the plantation with tunnels, and well-concealed fire bases. It was the latest of many such patrols for Nyberg and his seven-man crew, missions that were conducted over a surreal landscape that Joseph Conrad, the dark-visioned novelist, would have appreciated. Colonial capitalism flourished side by side with violent death that had started decades earlier in French-Indochina, when natives began trying to drive out the Western money-changers. "The plantation roads were always clear, you could see a long way," Nyberg said. "The roads were straight and well cared for, and off the road among the rubber trees, it was easy to get around because there was no underbrush to hinder the workers.

"You would say it was kind of unreal. They actually had rubber workers tapping the trees while killing was going on all around them. There were usually two or three workers in a group, usually barefoot. They hardly paid any attention to us, they had seen us so often and they simply went on tapping rubber trees and hanging the buckets to catch the dripping rubber sap. I didn't find this unusual. People had to eat; life goes on. This is the way it worked, the Viet Cong either waiting in ambush, or us going in among the trees to attack one of their fire bases that we had been tipped off about.

"The Michelin plantation continued to produce rubber just as if there was no war going on. I can only say that the plantation, the roads, the trees, everything, was very pristine. That's the only way I can describe it, pristine and untouched.

"The fighting on the plantation was continuous and heavy, and casualties on both sides were high, but there seemed to be no casualties among the thousands of rubber trees that were being harvested. I can't recall any of our artillery or chopper gunships being used to support us on the ground. If there were, then no doubt the trees would have been destroyed.

"I know there wasn't any support on September 27, 1968; it's a date I remember very well. We struck a land mine and our personnel carrier was blown over on its side. I was on top of the Green Dragon—that's what the Viet Cong called it—with five of my crew when we were hit, and we were all thrown into the air. My driver, who was inside the carrier, was killed instantly. My entire left side was injured, bones were broken, ribs, shoulder, and hip. I was evacuated within ten minutes, spent a month in the hospital, and then back into action."

While the Green Dragons of Nyberg's Second Mechanized Infantry Regiment were taking heavy casualties on one Michelin plantation northwest of Saigon, similar mechanized units of the 25th Infantry Division, also without air or artillery support, paid a heavy casualty price for over a month during earlier fighting at the Michelin plantation at Ben Cui, northeast of Saigon. Nyberg could not recall ever receiving such support, and let it go at that.

After completing his Vietnam tour of duty in 1970, Nyberg remained in the Army Active Reserve until 1995, retiring as a lieutenant colonel after serving as an ordinance officer during Operation Desert Storm. Besides his Purple Heart, he was awarded a Bronze Star and an Army Commendation Medal.

OVER THE FENCE

Frank Nollette was a "dorm rat" at CHAD High School from 1955–57, and one of thirteen members of the high school's third graduation class. He was the son of Frank C. Nollette, at that time a captain in the Quartermaster Corps stationed at the large Chinon Army Supply Depot. There was no high school at the Army Depot, so Nollette commuted by train and bus every Friday and Sunday between Chinon and CHAD, where he was quartered in a large Quonset at La Martinerie five days a week. It was during this time that Frank's father sternly admonished him to "Get out of the ghetto. Get out of the Golden Ghetto. Get away from it. Experience a real country." Nollette found himself posted to several Golden Ghettos during a twenty-eight year Air Force career, retiring as chief master sergeant on May 1, 1988.

With his family's lengthy military history, Nollette was a kid to the uniform born. His father, only five-foot four-inches tall, was the ideal ball turret gunner on a B-17 bomber during WWII. As an eight-year-old, Nollette had witnessed firsthand the effects of President Truman's decree integrating the Armed Forces. He enlisted in the Air Force three years after graduating from CHAD High School, and two years later he was a member of what was euphemistically called either Joint Services Combined Operations or Joint Services Special Ops. No matter what you call the group, it was never mentioned in news dispatches out of Southeast Asia, and that's exactly the way the U.S. government wanted it. The reason was simple: the group's missions were almost exclusively into Laos, where the world was led to believe the United States had no combatants. The unit was under the American Central Command in Saigon and operated out of Bangkok, Thailand.

"I wouldn't consider it an elite group because every mission was different, and the make-up and composition changed from mission to mission," Nollette said. "Sometimes we had Navy Seals, sometimes we had Marine Reconnaissance, sometimes we had Green Berets, and the reason I was involved was that I was good with radios. I was a good photographer, and they assumed with my name I spoke French [laughter]. Remember that in Cambodia, Laos, and Vietnam—all part of what was French-Indochina—almost everyone could understand French even if they couldn't speak it. French was taught in the schools. Because of this, and even in my limited way, I spoke more French in Southeast Asia than I did in France.

"Originally, I was in air crew protection. This involved the handling of everything for survival—from parachutes to safety equipment, gas masks, and everything else needed for crew survival. I also conducted a line of survival training.

"Everything we did was directed out of Saigon, where they had the power to pull in anybody they wanted for various missions. So we had people called in from literally everywhere for these short missions.

"Typically, we would put a team together and go in [to Laos] for about a week or a week and a half. We would get our job done, be debriefed in Saigon, and go home."

One of the missions was anything but typical. It began disastrously and got worse, with Nollette and his companions playing a deadly cat-and-mouse game with their Viet Cong pursuers as they dragged themselves through the jungles of Laos.

"In the parlance of the time, we were 'over the fence.' Our aircraft got hit by enemy ground fire. It was a totally unmarked C-47. We crash landed in the jungle. We had two people killed in the crash. We couldn't bring them out, so we buried them there in the jungle. We then deployed and hot-footed it to where we could get into friendly hands.

"The locals [Viet Cong] were in hot pursuit. We ran for two days. With the crew and everything, there were about ten of us. We were trying to leg it out to friendly territory when we decided to take a break in what was an old bomb crater.

"We got in there because it gave us protection and at the same time 360-degree visibility. All ten of us were in the crater, which was about twenty feet in diameter. We were pretty closely bunched. We knew that one well-placed mortar round would take us all out at one time. At dawn the next day, two grenades were thrown into the crater. Someone yelled 'grenade!' and we all piled out of there as fast as we could. I was up on the rim and started to roll over in order to get out of there when the grenades went off, hitting me in the head with shrapnel and dirt. I was knocked out. They bandaged my eyes and a special forces medical team treated me in the field.

"Basically I was a blind man being led through the jungle. We ran to our first designated pick-up point with the Viet Cong still on our tails. Nothing was happening there because the weather was so bad the choppers couldn't get in. We then headed to the second pick-up point, and again there was nobody there. So we continued heading east toward the Vietnam border. We were still over the fence [in Laos] at that time with the Viet Cong right behind us. After a day and a half of groping our way through the jungle, we finally got back into friendly hands, a Marine firebase on the border. We were there for twenty-four hours when the weather finally cleared and they were able to send in a chopper and fly us out to Saigon for debriefing. In Saigon, they took care of my injuries, basically facial contusions and that sort of thing. There was no permanent disfiguration. I was not hospitalized. But what had happened was well-documented, and I got a Purple Heart."

Gagné, Nyberg, and Nollette did not limit themselves to the easy minutia which is traditionally the mother's milk of countless high school memoirs. Nyberg did not dwell on his near-fatal motorcycle accident that crumpled his safety helmet and caused massive head and brain injuries, instead preferring to mention that he had been fighting a continuous and mostly successful battle to overcome memory lapses. Of great concern to him was whether the military career both he and his father before him had chosen was understood by those who had never been there. He never demanded that the sacrifices that he had made be appreciated, only that they be understood. His twenty-nine years of Army service was as an officer, and as such, he was a member of a privileged military caste he loved enough to readily accept without question his deployment to Iraq at the age of fifty. It came easily for him to say, "I loved it, the military was my life."

Nollette could easily be described as the Army's Boswell when it came to recalling the enormous changes that came after President Truman integrated the Armed Forces in 1948. Taking his father's advice to heart, he never forgot that each Golden Ghetto created by the U.S. military during the Cold War was merely a symbol, and that there was a much broader and enormously valuable world to be found outside the Ghetto's gates. It was a world that he and others like him in the military found worthwhile to protect, but simple, person-to-person loyalty was demanded.

"You were not supposed to let politics affect anything you were doing in the field. You worked under the commander-in-chief [the President], and there was a chain of command, and unless they gave you an unlawful order, you were expected to follow orders. You were there to do a job. There had to be a sense of loyalty not only to yourself, but to your buddies. You were more concerned about not letting your buddies down than anything else. Your loyalty, first and foremost, was to your buddies. They were protecting your back, and you knew it and they knew it. This was universally expressed by everybody I served with in the field. But there were a few who just didn't get it. If you were to talk to anyone who was in combat, and I don't care what war it was, the buddies that you had around you, they were your family. You didn't want to do anything to jeopardize it!"

There are numerous web links to the Châteauroux-Déols Air Station on the Internet. The CHAD High School Alumni Association would do well to consider Frank Nollette, Peter Nyberg, and Mike Gagné as charter members.

1964 – COLD WAR POTPOURRI – 1964

- On February 9th, the Beatles made their first American television appearance on the *Ed Sullivan Show* and their hit songs, "She Loves You" and "I Want to Hold Your Hand" sparked the Beatlemania craze.

- On July 2nd, President Lyndon Johnson signed the sweeping Civil Rights Act that made it illegal to racially discriminate in all public places and forbade employers to use race as a basis for hiring.

- On February 17th, General Maxwell Taylor, Chairman Joint Chiefs of Staff, and later, Defense Secretary Robert McNamara both disavowed any intention "at present" to send U. S. combat troops to Vietnam.

- Feeding growing U.S. anti-war sentiment, three blockbuster movies *Dr. Strangelove*, *Seven Days in May*, and *Fail Safe* were released, showing the military in an unfavorable light.

- On November 3rd, President Lyndon Johnson overwhelmingly defeated Senator Barry Goldwater in a bitter campaign that included LBJ's infamous "Daisy" TV ad, which depicted a little girl faced with a mushroom cloud.

- On October 12th, the Soviet Union launched the first space craft with a multi-Cosmonaut crew, who for the first time could move about their craft freely without space suits.

CHAPTER 14

OPPORTUNITIES, GOOD AND BAD

The French government . . . considers that the North Atlantic Treaty
Organization no longer corresponds to the conditions prevailing in the
world at present.

—March 8–10, 1966—NATO is told to leave France

From the outset, there was a clarion call of opportunity echoing from
the CHAD Golden Ghetto that resounded throughout a job hungry
France. Men and women responded. The pay was good, the workday
was relatively short by French standards, and even if you could not get
a job at the base, Uncle Sam's dollars streamed into the community,
creating a flood tide that tempted even the most dubious entrepreneurs.

Marguerite was eighty-two, or perhaps eighty-three, when I found
her. I had been trying to contact Marguerite for three years. Her career
as a prostitute had taken on almost mythical proportions, and her du-
rability had become legendary. Only after it was agreed not to mention
her surname or home address did she guardedly agree to tell her story.
She surfaced during a chance encounter in Brassioux that occurred
during the village's fiftieth anniversary celebration. A woman looking like a young grandmother willingly engaged my researcher Valerie
Prôt in a revealing conversation, although they had never met before. It
was eerie that our search for Marguerite was virtually a mirror image
of our efforts to find Irene Smith.

As Valerie observed, Marguerite was obviously one who did not
take public appearances lightly. At the Brassioux celebration, she wore
a white denim jacket, brown trousers, modern but classy trainers, and
a pair of sunglasses with light blue lenses and a trendy geometrical
shape. "She is short-haired and has a good style of makeup, nothing
too flashy," Valerie said. "Perfect harmony of the whole."

Marguerite's story is one in which the bizarreness threatens to
crowd out the sad, underlying realities. Marguerite was twenty-six

when she arrived in Châteauroux in 1951 from Nice. Her mother told her that a big American air base was being built in Berry, and there would probably be plenty of jobs. In a succession of emails, Valerie takes up her story from there.

"I told her about the book you were writing, and she explained that actually she had had nine children. She kept five of them, and she gave the four others for adoption to American ladies the day they were born. She said the men who fathered her babies were basically thugs and should have been in jail. One of her babies was born in the winter, and when she presented the infant to the GI father, he disdainfully abandoned the baby in the snow. She told me that she had been beaten by her boyfriends. She went to the MPs to complain when she was beaten and when she was bleeding, but nothing was done."

When Marguerite arrived in 1951, she had no job skills, money, or prospects. What followed was a long succession of American airmen with cash in their pockets and promises to break, according to Valerie. "Marguerite added that the GI fathers had promised her that when they returned to the States they would send her money. They never did. She said that she could write a book, but she had five children who despised her because they didn't know their father—I mean, five different fathers."

Valerie discovered that Marguerite lived in a fashionable, high security apartment building in downtown Châteauroux. She attempted to rekindle the conversation begun a fortnight earlier in Brassioux, but found Marguerite curt and dismissive, explaining that she was very tired and ill. A shopkeeper friend of Valerie's had spotted her speaking into the apartment intercom and expressed surprise that she knew anyone in the neighborhood.

"I told her about the woman, and of course, she sees her every day. She knows that she is a prostitute. Marguerite had a boyfriend who died not long ago. She had stopped her activity for a while, but since he died, apparently she is back to business. My friend told me that she has seen her with various men. She guessed that Marguerite was unfriendly over the intercom because she had a client with her."

One could easily ask why an octogenarian prostitute's story is compelling enough to be included in this book. Simply stated, it reveals some ugly truths about military occupation no matter how outwardly friendly it might be. Much has already been written about the dozens of young prostitutes who arrived in Châteauroux every payday, remaining until the GIs' money ran out. Then it was back to Paris and Toulouse, before returning in two weeks to prime the pump again. The success of Châteauroux's hostess bars could be gauged by the number of smiling women the owners let hang out, either paid as so-called students who were there to earn their college tuition or as freelancers. Everyone seemed to be making money.

This was the world in which Marguerite functioned. She had nine babies during the sixteen years leading up to the drastic downsizing of CHAD. During that time, Châteauroux and the surrounding areas had become, in effect, a sprawling garrison town that could hardly be called synergistic. Locals mistakenly believed that the base could not exist without them, when the opposite was true. It would be hard to find an officer or an enlisted airman who did not believe that Châteauroux was a one-way street with the U.S. military taking and the locals giving.

Looking for a means of support, Marguerite found herself in a world where child abandonment was all too common. In Châteauroux, she competed with others, including the infamous Nine-Fingers. Marguerite, for reasons only she could fathom, had never perfected a means of escape. Instead, she endured beatings, one child after another, and broken promises, with no remorse. If there ever had been an opportunity for Marguerite, the unanswered question must be "where did it go wrong for this woman?"

Evidence indicates that it is doubtful that Marguerite had ever set her economic sights very high, and with the opportunity offered by the profession she had chosen, social acceptance would be dubious at best.

For Others, the System Worked

René Coté was a staff sergeant in the base JAG office, a position that provided him with security clearances. As a result, there were privileges that provided him with money-making opportunities at the base which were not open to his competitors. His security clearance continued for one year after his discharge, and as a result of his marriage to a French woman, he also had *carte de séjour*, giving him permanent residence status. Thus armed, Coté was ready to take the first steps in creating a tidy marketing empire that extended from Berry to Paris.

"I was out of work and looking for a job wherever I could," Coté said. "I went up to Paris and started nosing around, investigating the same way I did while with the JAG office. It was in the bar at one of the best hotels in Paris, the Georges V, when I heard a man with a Brooklyn accent leading a conversation at one of the tables. His name was Charlie Fighth, a real operator, a dealer, a real hustler. He was looking for a salesman to sell cars, American cars, in France if not throughout Europe. I went over to the table and told him I could do the job. Twenty minutes later I had a job."

Coté went on to describe a frenetic series of events that occurred that same day in 1956. He was given a Ford Victoria to drive to Châteauroux followed closely behind by a Mercury, also supplied by Fighth, to be sold. "I got another Mercury and after that came along another Ford Victoria, all of them coming down from Paris on that same day," Coté said.

That was only the beginning. With his security clearance, he was able to display eight cars at a time along La Martinerie's major thoroughfares. Coté also operated two car rental offices, one downtown and the other at the La Martinerie PX. At the same time, he maintained a sales office in downtown Châteauroux with up to twenty-seven new cars on display. The favorite among GIs was the Renault Dauphine, a small, very tight, four-seat sedan. Coté sold it for $1,000, with $195 down and an eighteen-month contract. "The Renault dealership in Châteauroux was selling 450 cars per year, and after one year I was selling more than that," Coté said.

Coté described his wheeling-and-dealing techniques. He would sell a car to a GI, repurchase it, and then resell it to the French at even a bigger profit than the original sale. This system worked well as the base was closing. Another coup included an agreement with the local Renault dealer that there would be no competition by the dealer on the sales turf Coté had established, and in return, Coté would send the cars he sold to the dealer for all warranty work and repairs. After the base closed, Coté began selling hard-to-find luxury cars like Jaguars, BMWs, and sports cars from a garage not far from the old base. Auto sales and resales secured Coté's bona fides as an operator, but his leap into the insurance universe demonstrated a pure opportunist at work.

It was easily accepted that Coté would offer all kinds of coverage, at what he claimed were the best rates possible. But it was his former position as an NCO in the JAG office that provided him with insider data on how another insurance market could be manipulated: babies.

"I had access to the base hospital records as far as births were concerned. The telephone number would have the name of the parents and I would call them," Coté said.

"I had three things working for me at the same time. I would call them on the phone, set up an appointment and then make a proposal. The first offering would be for baby insurance, and then possibly selling them a car which they probably would need with a new baby, then selling them insurance on the car. The baby insurance was called the PIT plan. This would put enough money in place for the child to get an education, and the money would be waiting for them when they were twenty years old."

Coté knew from the very beginning that old fashioned, hard cash was the currency of the realm at the Golden Ghetto and he went out to get it. There are two photographs that clearly show that Abel d'Aquembronne and Coté were kindred spirits, although they had never met. In one photo, a seemingly endless row of modern red and white buses were parked along a Châteauroux street, with their smiling drivers proudly standing nearby. The second photo was taken, fittingly enough, outside of Saint-André, Châteauroux's largest church. There were six vehicles parked near the church—three ambulances and three hearses. As in the first photo, smiling drivers and attendants were standing beside their vehicles. They were testimony to d'Aquembronne's boundless enthusiasm that never diminished after his original bus company was put out of business by Germans during WWII, and was resurrected when the Golden Ghetto offered opportunities to a guy who arrived in town with a single, broken-down taxi.

Antoine Price, who married d'Aquembronne's granddaughter, Veronique, spent many hours talking with his grandfather-in-law during his later years. They were comfortable together despite the vast difference in their ages, and even developed a relationship that enabled them to share a bottle of Suze, an old-recipe, quinine-tinged elixir that the old man loved.

"He first started with only one car," Antoine said. "It was not a quality car, because he told me that in the beginning, half way through an average taxi trip, he would tell his passengers to get out of the taxi because he would have to work on it to make sure that it would keep running. He had to remove the backseat because from what I could understand, he had continual, different problems that had to be fixed, and that was the only way he could get at it. He would have his passengers sitting on the side of the road while he fixed it. That's how he started, that was the beginning.

"Before coming to Châteauroux, he had been successful, but WWII and the Germans changed that. During the war he had buses, and he had to hide them in nearby farms, in barns where they couldn't be seen because the Germans were requisitioning everything. They especially needed transport vehicles. They eventually took his. From what I understand, he came down from Paris with only one car, the one that I just described."

D'Aquembronne's daughter, Marie Antoinette, was a teenager during the CHAD golden years. Her father's buses were a symbol of everything good that she remembered. "I met my first American on one of my father's buses. They would get off the buses at night to check out the town, see how it was, and have a drink away from the base. There were two bus lines that transported workers on the Châteauroux-La Martinerie line, and French people who worked in offices. He also had taxis, excursion buses, ambulances, and hearses."

In his talks with d'Aquembronne, it became obvious to Antoine Price that the old man had never missed an opportunity to make an extra buck. Renting a hearse was okay, as far as it went, but it was the complete funeral that offered the most possibilities. "He had a warehouse in a small town outside Châteauroux where he stored funeral equipment. You know, holy water fonts, and also those little hand sprinklers, I don't know what they were called, that people dipped in the holy water and sprinkled on the caskets."

His contract with the American military was a sweetheart deal. It included all bus transportation between the base and the 410 housing area and downtown Châteauroux. He eventually sold the company, and that company is still in existence today. It is called the *Transport de l'Indre*.

D'Aquembronne sold out, but didn't go down without fighting. His daughter recalls the dismal economic scene after the base closed, a situation of which her father tried to make the best: "My father continued to transport people, but by this time it was the French. Almost all of the Americans and their families had left from Brassioux and from 410 housing. The French Army had taken over the base. There were still some skilled French workers at the Déols air base, my father continued to transport them. And then the jobs were lost, and all those highly skilled airplane repairmen went to Toulouse. It was all over."

Marie Antoinette, as an only child, inherited her father's fortune and over the years acquired restaurants, discos, a chateau in Ste. Colombe, part of which has been converted into an expensive *chambre d'hôte*, took over her father's small villa in Ronce les Bains on the Atlantic coast, and now owns an apartment in Châteauroux.

During his final years, Abel d'Aquembronne lived alone with two dogs in a small house in downtown Châteauroux. "The only time he went out was to go to lunch at one of his daughter's restaurants, where they would fix him up with a meal. It was pretty much a daily thing for him," Antoine Price said. Abel, described as a tall, gangly man, had contracted inoperable intestinal cancer and was allowed to die peaceably. He was among the first entrepreneurs to fully appreciate and take advantage of the opportunities offered by the Golden Ghetto, but he was not the last.

Pride—Wounded and Restored

Maurice Garnier was seventy-five when interviewed in 2006, and he explained the career-changing decision made easy by a 'good old boy' network that came to his aid. I learned that job hunting in the United

States and France—especially for the good jobs—is not that different after all.

"I had come to the end of my tour as an officer in the French Army. I had choices. One of them of course was to remain in the army, but I had become interested in transportation and supply. My decision was to leave the army. If you've served in the French military, you'll always have friends who will try to help you out. They told me I might be interested in working at the air base in Châteauroux—certainly that's transportation.

"I was the director of French personnel working at the base, assigning jobs at three warehouse complexes, huge storage areas for material shipped to air bases in Europe and North Africa. I had a sizeable office staff that got bigger as the base grew. We needed a good number of office people to handle the paperwork. Most of them were young girls."

A heavy premium was placed on hiring clerical workers; experienced or freshly out of school, it made little difference. As the supplies piled up in the warehouses, so did the paperwork in the offices, and in triplicate. Young French women, many of them fresh from the farm and with only the most rudimentary office skills, were hired by the dozen. It was the best-paying on-the-job learning experience they could have imagined. And as Garnier learned, parental pride often took a beating.

"When the base opened, the decision was made not to undercut local employers by paying much higher wages than they could afford. A standard was set. Roughly, it was decided to hold that starting salary scale at ten percent over what was paid for the same civilian job. Of course when the experience went up, especially fluency in English, the salary went up.

"The effects of the nice wages could be seen differently among the young office workers. Many of them were girls right out of school, and this was their first job. They had only the most basic office skills, but it made no difference because the demand for clerical workers was very high. Maybe I can explain it best by telling you about this young woman who handled really basic clerical duties. She had just received her first paycheck from the American military and taken it home with a great feeling of pride. I saw her in the office the next day. She wasn't happy and I asked her why.

"She was only eighteen. She came from a working class family. She explained that she placed the check on the kitchen table and when her father and mother saw it they were shocked. They didn't say anything. They just stared at the check. She told me her father was a laborer who worked hard all his life. I got to know who the father was and he wasn't at the bottom of the pay scale for laborers. The piece of paper on the table was for more money than any paycheck he had ever received in his life. You can only guess what went through the mind of the other family members when they realized that this young kid might have become the breadwinner of the family.

"Her story was similar to others that I had heard about. There was good money to be made at the base, and for those who couldn't get in on it there was a lot of jealousy and in this case, hurt pride."

If a paycheck from Uncle Sam could diminish the self-worth of a laboring class father, it could also restore the pride of technicians who had had the tools of their trade stripped away. Maurice Juanot was not a stranger to fear and anxiety. They had become old friends, and the air base at La Martinerie was the reason. It was hard to keep track of the base's status as it changed hands among first the French, then the Germans, then back to the French, finally ending up with the Americans in 1951. La Martinerie, hailed during World War I as the best pilot training facility in France, suffered unthinkable indignity in the hands of the Germans.

Juanot a skilled aircraft technician explained it this way: "During the war, the Germans who occupied the aeronautical shops in Déols used them for other means. They built pots and pans and suitcases! Also at that time, the Luftwaffe destroyed all war material belonging to the French Army on this site. They moved the Bloch-152 fighter plane to Germany for pilot training."

After the war the French reopened the aeronautical shops in Déols, and Juanot went back to work at a job he loved, but he was never sure how long it would last.

"The workshop had less and less work and we were getting low on tools, so when the GIs arrived in 1952 we were relieved to keep our jobs and to learn the latest aircraft technologies that they brought us. Before arriving in France, the American GI had been given a negative impression about French workers. American officers and engineers soon saw that we were professionals.

"They would often say, 'Let the French do their job and you do yours.' They would make us work apart from each other because we had different work habits. Anyway, we didn't know the planes as well as they did, but we did know how to work and work well.

"For example, I remember GIs working on a plane called the Canberra. The work we did on it was on the day shift and they would tear it down on the night shift! This lasted about three months until the commanding officer said that the situation had lasted too long, let the French finish it. When we got the plane to ourselves, the work was done in a few days and the plane was delivered. No one likes to be shown up, and as you can see from this, working with the GIs wasn't always easy."

Barely perceptible, a wry, perhaps even smug smile crossed Jaunot's face as he recounted the Canberra episode. And why not? The Canberra was not an American plane, but British. The twin-engine, multipurpose Canberra was fast, maneuverable and performed remarkably well at high altitudes, all characteristics for a potential NATO "spy plane." Complex overhauling would be needed: a longer fuselage to house a gaggle of cameras, extra fuel tank, and electronics.

American Air Force technicians struggled for weeks, only to have their boss call them off the job and turn it over to the French to get it successfully completed. Jaunot's wry smile was vindication for all those years when a war had taken the tools of his trade away from him. The Americans had given him his tools back and he seemed to be saying that he and his coworkers had something to prove and they were doing just that.

Juanot's anecdote about the Canberra could very well sum up a learning process that would go on for sixteen years, a journey of discovery for both French and Americans.

L'enfant Terrible du Cinéma Français

Gérard Depardieu was born in Châteauroux on December 27, 1948. As a rough hewn, street-smart kid, he was not yet a teenager when he recognized the opportunities the air base could offer if he played his cards right. "During my youth, what I consider to be the first half of my life, I wanted to re-do my family," Depardieu was quoted by Paul Chutkow in his book, *Depardieu, A Biography*, cited extensively in a *Paris Match* magazine article from March 3, 1994. The magazine described how Gérard and his older brother Alain ingratiated themselves to the Americans, especially those who called Joe's From Maine a home-away-from-home, and eventually the two brothers would "enjoy a special stature: they are at the same time buddies, mascots, interpreters and mediators." Depardieu quickly realized that if he were to continue traveling in the exalted circle offered by the young Americans, it would take money and at least casual friendships with GIs that would give him access to the bountiful largess that could only be found at CHAD.

Big, strong, and oversized, it was hard not to notice and remember Depardieu when he accompanied his newfound GI friends onto the base to roller skate, bowl, or enjoy hamburgers, French fries and Coke at the snack bar. It had been more than forty years since Pat Thacker had enjoyed the good life at CHAD High School, but she remembered with clarity the times she had seen the teenaged Depardieu sauntering about the base in jeans and a tight t-shirt. "He used to come onto the base. I never had any personal contact with him, but I did see him around," said Thacker, who, like everyone else, did not know she was looking at a kid who would eventually become one of France's reigning superstars. That realization did not materialize until decades later. "I was surprised to find that a big movie star, such as Gérard Depardieu, was from the region."

Roller skating and bowling were a lot of fun, but they did not represent the targets of opportunity Depardieu was looking for. "I entered [the PX] with adult Americans, from whom I bought their ration cards. They had some for all their things. I told them: 'You don't drink, you don't smoke, I'll buy your card.' I was very young. It was my little deal," said Depardieu, as quoted by Chutkow. *Paris Match* explained how profits from Depardieu's petty black marketeering were used. "He re-sells the whisky or cigarettes [that he gets with the ration cards] to the French to earn money to pay for his jeans, his t-shirts, his burgers and his French fries."

Depardieu was thirteen years old and attending Saint-Denis school when he and his buddy, Serge Dubreucq, developed a scheme to make some extra pocket money during an annual church charity drive. It involved the sale of packets of stamps for two francs each. "The GIs or their wives gave us five or ten francs for a sleeve [packet] that was worth two francs, we kept the difference," Serge said. "We bought the sleeves from the other children, and asked Father Durand to give us more to sell. That permitted us to make a little spending money."

Depardieu called it quits at Saint-Denis after he was accused of theft from a school donation box. "Someone accused me of having stolen money," says Gérard. "Father Lucas called me in. He closed the cur-tains, and then, pof, pof, he slapped me two times. God, I hated him. Then, he united all classes, in the yard, and he said: 'He is in quarantine. Anyone who speaks to him will be punished.' That was the worst shock, total humiliation. Because I was punished for something I didn't do." Depardieu said that he found out later that year who the thief was, con-fronted him and got him to admit the theft, but as a matter of principle, he would not turn him in. "I took that to heart," he said. "That vacci-nated me against school."

Doubtless, Depardieu carried with him the bitterness of the false theft accusation when he pulled out of Châteauroux for the Riviera resort of Cap D'Antibes. He quickly made friends with the shy Michel Pilorgé, who had enrolled in the Theater Nationale Populaire in Paris and invited Gérard to come along. Depardieu impressed theater in-structors and was allowed to attend classes without paying tuition. He recalled the first time he got on stage to perform, "I went up on stage, in the light, and there, I had all my background in Châteauroux. It was like being in front of the cops. The only way to get out of it was to smile, and the smile became a giggle inside myself. I didn't say a word, but they have all said I smiled stupidly; then I began to laugh hysterically, and the whole room began to laugh. I did not even have to speak."

Sheila Trubacek could hardly be called the typical "Army brat." Her father, John Trubacek Jr., was chief of special services for the Army in Europe, and as such, was in charge of scheduling visits by luminaries such as Joe DiMaggio and football star Bob Waterfield. This gave the beautiful high school cheerleader practice with how to behave around celebrities. Sheila was tall and athletic, and after graduation from the Orleans Army Base high school in 1957, she went to Paris to pursue a ballet career that was quickly aborted. "I was so very tall, I knew it wasn't for me. I am almost five feet, ten inches tall, when you're wearing toe shoes in a ballet, the ballerina cannot be taller than the male dancer," Sheila said. Modeling came next, and in what better place than Paris, where Sheila fashioned a twenty-nine year career as a top runway model for Coco Chanel, Yves St. Laurent, and Christian Dior. "I was Coco's favorite. She gave me one of her dresses and shoes one year as a Christmas present," Sheila said.

Interviewed while at the home of her daughter, Deborah Lynne Templeton, in Virginia Beach, Virginia, Sheila displayed a wonderful faculty for remembering the celebrities with whom she partied. Sheila said that among them were French movie star Jeanne Moreau, often accompanied by a young, very large, provincial, aspiring actor, Gérard Depardieu.

"Depardieu was very rough around the edges, and Jeanne Moreau took him under her wing as I guess you could say her protégé," Sheila said. "He was what you would call *l'enfant terrible*, you know, kind of a bad boy but not really bad, just rough. Jeanne smoothed him out, taught him how to behave in this new kind of company for him, and how to dress, all those important little things."

There were return visits to his hometown of Châteauroux by Depardieu, who was no longer the same embittered, falsely accused teenage thief that had left town in 1963. It was easy for author Chutkow and *Paris Match* magazine to paint a picture of Depardieu in broad strokes, but the people interviewed in Châteauroux were much more parochial.

"He was a good customer. When he got started as an actor it was at a studio in Paris. But he used to come back down to Châteauroux every week," said Joe Gagné. "We used to talk a lot. I remember one time when another customer, an American, was looking at him. He asked him, 'why are you looking at me?' The guy didn't know what to say."

"He was a good guy, a good actor. But he was the kind of a guy you either liked or didn't like," Châteauroux Mayor Jean-François Mayet said. "He didn't like Americans. He liked to fight and he was always fighting Americans [Laughter]. For him fighting was fun, he was always laughing when he was fighting."

"He was a troublemaker. He did have a big mouth. He thought he was a big shot, but he never was. My father threw him out just about every night he came in. But he'd come right back again the following night. He was a big mouth," said Mike Gagné.

"He had a moped," Fernand "Gypsy" Marien said. "We used to call him 'Vroom-Vroom' because his moped had no muffler."

"Most of the time he didn't even have any money in his pocket when he came in here [Joe's From Maine]," Annette Gagné said. "My mother said he was a trouble-maker, a loud-mouth. My dad was tired of having him around, so he threw him out. It's no wonder that when he was making a movie nearby, he never stopped by to say hello."

"He came in several times to the nightclub, the Nicholas II, which I was running for my mother-in-law, Marie Antoinette d'Aquembronne. He was usually pretty drunk," Antoine Price said. "He came in with the restaurant owner that he palled around with, Monsieur Bardet. It was annoying that they would come in and drink and then they didn't want to pay for their drinks. When they were leaving, Depardieu would turn to my staff as he was walking out the door and tell them, 'My presence is all the payment you need.' He didn't even give me the option of offering him a drink or two, which I would have done. The bill eventually got paid, not by Depardieu, but by Bardet."

Of course, there were other opportunists—although, with one possible exception, they were hardly in the class of Coté, d'Aquembronne, and Depardieu. Joe Gagné became a mythical figure in CHAD lore by meeting an insatiable demand for hamburgers, BLTs, ham and cheese sandwiches, chili and French fries. His cash flow grew to the extent that he had opened a bank account in Geneva, Switzerland, and the money was still intact after the base had closed. Simone Nickles, a former NATO employee at CHAD and an old friend of Gagné's, was working in Geneva when he knocked at the door of her apartment. "He said he had money in a Swiss account and didn't know exactly how to get it out," Simone said. "He asked me if I could help him. Let's just say I did the best I could and leave it at that."

And then there were the young ladies from Paris who found work in Châteauroux's hostess bars insisting they were university students, only working there to earn tuition. Owners of the region's mushroom caves had never had it so good; the fleshy fungi they were offering for sale were a favorite ingredient in the recipes of those Air Force officers' wives who could pull themselves away from the Officers' Club to do their own cooking. Wineries stretching as far as the outer reaches of Sud Loire Department were discovered with a vengeance. Who needed the pricey Bordeaux wines like Margaux, Saint-Emilion, St. Estephe, Graves, and Pomerol when just around the corner you could select a Chinon, Bourgueil, Cheverny, Reuilly, Saumur, Valencay, or Muscadet at half the price? Business was booming for central France's vintners. It was hard to believe that this could all be coming to an end.

THE HAMMER DROPS

All doubts were dispelled during these fateful days in March 1966, when from the eighth through to the tenth a series of memoranda proclaimed Charles de Gaulle's intentions. The French proclamation, part of which is quoted above, declared that the NATO treaty was in effect moot, and went on to say that "the French government considers that these agreements in their entirety no longer correspond to the present conditions. Which lead it to reassume full exercise of its sovereignty on France's territory, in others words, to accept no longer that any foreign units, installations, or bases in France be responsible in any respect whatsoever to authorities other than the French authorities."

A deadline of April 1, 1967 was set for removal of all foreign troops, and the ink on the proclamation was hardly dry before preparation for the abandonment of CHAD were under way. In his book, *The Last Great Frenchman*, published in 1993, British author Charles Williams observed that de Gaulle's ministers had not been consulted and panic reigned throughout the ministerial offices in Paris. Typically, it was a personal, "hands-on," and provocative decision by de Gaulle, as witnessed by the letters he sent to the four most powerful Heads of State among the other NATO members. The handwritten messages were essentially the same, but with subtle differences. The message to British Prime Minister Harold Wilson emphasized his fear that France would be drawn into the Vietnam War. A slip of the pen in his message to German Chancellor Ludwig Erhard caused concern when it mistakenly referred to the 1869 Franco-German treaty and not to current concerns, and this left German officials shaking their heads with dismay. Other messages went to Italian President Giuseppe Saragat and President Lyndon Johnson, with de Gaulle assuring LBJ that France would remain in the alliance and "when the time comes, a party to the Treaty." This assurance was not included in the other three messages.

U.S. Secretary of State Dean Rusk and Defense Secretary McGeorge Bundy expressed their anger when they went before Congress, and Bundy was strongly critical of de Gaulle's decision. Nevertheless, on March 29, 1966 it was announced that French military staff would leave Allied command on July 1, and that all foreign troops would be out of France by the following April. NATO was no longer welcome and its headquarters would be moved to Brussels.

The Vietnam War was grabbing all of the headlines in the United States and by comparison, coverage of America's forced departure from France went virtually unreported. However, an occasional news item would appear that questioned which country was being hurt most by de Gaulle's calculated actions. The closure of CHAD was the focus of only a few articles to appear in the American press, this one on September 19, 1966, when the author wrote: "Châteauroux is one place where the shutdown of an American base is hurting the local French economy more than it's hurting the U.S. The French population in and around Châteauroux are watching three-fifths of the local economy move out with the Americans. Earlier this summer, nearly 700 French workers who had jobs repairing planes at the U.S. base demonstrated against the shutdown. French riot police broke up the demonstration."

This straightforward assessment in a U.S. newspaper of how Berry would suffer the economic consequences of de Gaulle's decision was a rarity. More often than not, the American press were bemused at what they considered de Gaulle's outlandish act. *The Baltimore Sun's* cartoonist Yardley depicted a tall man with a hawk-like nose pointed snobbishly upward, wearing an apron from which hung a ring of keys. The cartoon was titled "The Adamant Concierge." With one extended arm, the concierge pointed to a NATO officer with a "check-out time" sign sitting on the hotel registration desk clearly visible under his arm. Yardley could not resist making a further spoof of the entire episode, inserting the legendary GI graffiti, "Kilroy was here." The concierge's other hand held a newspaper whose headline read "Deadline for Paris Control of Allied Troops in France." While the de Gaulle-like concierge was issuing his eviction notice, two cats were eyeing each other from their vantage points on the floor. One of them was wearing a black beret and smoking a cigarette while winking lasciviously at the other feline peering from behind the NATO officer's leg. *The Baltimore Sun's* reaction to de Gaulle's edict was typical among many news outlets in the United States—that it was hard to take seriously de Gaulle's attempt to restore *Grandeur de la France*.

De Gaulle's animus toward the United States had been evident for years and was vehemently expressed in 1966 when he described the Vietnam conflict as "the bombing of a small people by a very large one." A couple months later, he pulled out all the stops, stating Vietnam was an "unjust war, since it results from the armed intervention of the United States on the territory of Vietnam, a detestable war since it leads a great nation to ravage a small one." Meanwhile, technicians and salvage crews were breaking down U.S. military bases in France and shipping them in bits and pieces to other NATO installations. Bundy, Rusk, and the Air Force carefully managed news of the American pullout from France, downplaying that NATO's tactical airpower in Europe would be decreased. But nobody in France was fooled.

1965 – COLD WAR POTPOURRI – 1965

- On November 26th, the *Astérix* became the first French satellite to be put into orbit, launched by the Diamant (diamond), the first space launching system not designed by the United States or the Soviet Union.

- On April 27th, iconic radio and TV commentator Edward R. Murrow, a lifetime heavy smoker of up to seventy cigarettes a day, died of cancer. Four months later, cigarette ads are banned from United Kingdom TV.

- On December 9th, *A Charlie Brown Christmas* based on the "Peanuts" comic strip, aired on CBS for the first time, earning Emmy and Peabody Awards, and surprising network bosses who had feared it would be a flop.

- On March 7th, the first of three Alabama civil rights marches from Selma to Montgomery was dubbed "Bloody Sunday" when marchers were attacked by state and local police using billy clubs and tear gas.

- On July 30th, a momentous bill signing week for President Johnson began with a bill increasing Social Security benefits, another creating Medicare, and on August 6th the Voting Rights Act became law.

- On October 17th, Richard Nixon responded to growing anti-war protests stating, "We must never forget that if the war in Vietnam is lost . . . the right of free speech will be extinguished throughout the world."

CHAPTER 15

ABANDONMENT

A fécondé soudain ma mémoire fertile . . . la forme d'une
ville change plus vite, hélas! que le coeur d'un mortel.
—Charles Baudelaire, "Le Cygne à Victor Hugo"

Throughout the five years spent researching, interviewing, and visiting Cold War landmarks throughout Berry, it became evident from the start that this book would not adhere to any generally accepted formula of creative non-fiction. Based in large part on memories fifty–sixty years old, *Golden Ghetto* is subjective to the extreme. Reaction to why an American would even be writing a book like *Golden Ghetto* ranged from pleasant surprise to innate French suspicion about my motives; from dismay caused by faulty memory, to joy because I had shaken loose recollections that had been dormant for decades.

This final chapter is a kaleidoscope of tumbling and sometimes colliding perceptions that were still open to interpretation and contradiction almost a half-century after the Stars and Stripes came down at CHAD and the gate was locked.

When poet Charles Baudelaire wrote "The Swan" in 1857, it had been nine years since he had earned the right to pen the line quoted above: "Suddenly made fruitful my teeming memory . . . the form of a city changes more quickly, alas! than the human heart." Baudelaire had fought on the barricades in Paris in 1848 during the *journées de février* and the *journées de juin*, and in 1851 was among those who resisted the military coup of Napoléon III. The revolution failed, and the Paris that Baudelaire loved and fought for would be forever gone. His beloved *Place de Carrousel* with its outdoor bookstalls, fishmongers, prostitutes, cafés and cheap wine bars, a pond that dozens of swans called home, beggars, sidewalk vendors hawking fresh produce and questionable meat had disappeared. Agonizing over this loss, the poem's last surviving swan, its pond paved over with cement and cobblestone, wanders hopelessly through the dust of the new *Place de Carrousel* created by Baron George-Eugène Haussmann. *"Je pense . . . à quiconque a perdu ce qui ne se retrouve jamais, jamais!* (I think . . . of whomever has lost that which is never found again, never!)," Baudelaire intoned.

Haussmann was commissioned by Napoléon III to transform Paris by destroying old neighborhoods, relocating the poor laboring class, and replacing it all with the wide, grey-faced boulevards constructed above an underground sewage and sanitation system that we know today.

It had been almost a century since Baudelaire's forlorn swan wandered homeless in Paris, a sad tribute to progress. The swan and its peers were ancillary victims of the urban upheaval Napoléon III had ordered and Haussmann had engineered. Fear, real or imagined, provided the blueprint for Haussmann's grand design for Paris. Paris was no longer the medieval city that Baudelaire loved, and its people were transformed from citizens to inhabitants.

In 1951, it had been five years since the Iron Curtain had descended across Europe and Cold War fears had had to be assuaged. There must never be a repeat of the war that had ravaged Europe less than a decade earlier. Yankee dollars kept flowing in, and until Charles de Gaulle's edicts on March 8 and 10 in 1966, few believed the stream of greenbacks would ever end. Prosperity, made possible by Uncle Sam's big pocketbook, proved to be the only development about which recollections converged. During those early months of 1951, the scope of what would eventually transform Berry challenged the acuity of its citizens. It was hard to believe what was happening.

At the height of its commitment, the United States had twenty Air Force and Army bases throughout France, spending hundreds of millions of U.S. taxpayer dollars for their construction, equipment and maintenance. The Châteauroux-Déols Air Station was the largest support depot in Europe supplying men and material to far flung NATO bases on three continents. It was unthinkable that CHAD would ever close.

Very few French, who had seen their lives economically and socially transformed by the CHAD Golden Ghetto, wanted to believe the danger signals from Paris. Stunned reality set in when they learned of de Gaulle's imperious edict. Conversely, the American military never deluded itself about de Gaulle's intentions. "We all knew that it was inevitable. We didn't know exactly when the hammer would fall," said Bob Goggin, "but everyone knew it was coming. We showed a great deal of deference toward the French. We got along very well with the military and with the French authorities locally. But we also knew it wasn't the same thing with de Gaulle in Paris."

Monique Coté remembered that the smell of betrayal had filtered across the miles from Paris, and that it was sensed by everyone regardless of his or her politics. It was as offensive to the Communists as it was to political conservatives on the Right. "The Commies had never really been that bad toward the Americans. There were only a few demonstrations, but they weren't that bad," Monique said. "There were big demonstrations when the base was closing. But these were not demonstrations saying "Yankee, Go Home." They were shouting: *Ce qui est eronné avec de Gaulle!* (what's wrong with de Gaulle!) *Vous ne pouvez pas faire confiance á oncle Charlie!* (You can't trust Uncle Charlie!)"

Maurice Jaunot, a flight line technician at Déols air base, had experienced elation when the U.S. Air Force arrived and gave him a steady, good-paying job for more than a decade. His response to news that the base was to be closed was painfully visceral when he asked, "What is going to become of us?"

Ironically, the high salaries paid to skilled technicians like Jaunot created an economic problem in which many French companies did not want to get involved. Louis Morin, the owner of a Levroux building supply company, insisted that there were indications, even when CHAD was open, that commercial and industrial growth in the area would suffer because there was no trained help available. "Many people forget that there had been plans for an industrial and commercial complex, but it never happened because all of the good workers were employed at the base. The plan never materialized."

Fernand "Gypsy" Marien shared Morin's opinion. "It was very difficult because at the time of the base, there were plants like Michelin, important firms which didn't want to settle in Châteauroux because there was not [a large] enough work force. People were employed by the Americans. As a result, these factories settled in Clermont-Ferrand, Montlucon. So after the departure, there was a sort of gap concerning the work force."

MEMORIES—GOOD AND BAD

Among my earliest interviews were the recollections of two of the region's powerbrokers. "It was one big show, day after day. It was like standing along the road and watching a circus go by," Châteauroux Mayor Jean-François Mayet said in 2005, describing the impression that the arrival of the Americans made on him as a young boy. "We had never seen equipment like this before. It was so big we were awestruck." For Mayet and Senator François Gerbaud, it was the memory of what was happening despite the elements that most readily came to mind. For Gerbaud, the enormity of everything as it appeared to him as a young boy conveyed wonder and danger. "The base setting up was incredible," Gerbaud said. "I remember one day, a crane that to my young eyes was enormous, it had lifted and was carrying a large electrical transformer. The muddy ground was not solid enough to support it. I remember workers putting slabs of concrete on the ground with the transformer hanging above their heads. To me it appeared very dangerous, almost courageous."

Leandre Boizeau, publisher of *La Bouinotte* and a happily prosperous Communist, also heard the noisy rumble of heavy construction equipment, but his introduction to a lawn game favored by America's elite was among his most cherished memories. Badminton provided the social glue that enabled a young Boizeau to share a friendship with a U.S. military brat. One day, Boizeau would be handing out anti-American pamphlets on a Déols street corner, and the next he would be whacking a feathered birdie over a badminton net at his new friend's home. "This guy, whose photo I have, had set up in front of his house a badminton game which I didn't know about at all. So the two of us used to play badminton," Boizeau said. "We used to play and I remember that he hated to lose, he really hated it."

The irony of the two kids' friendship was implicit. The Communist parents of one hated the American presence in Berry, while the other was only there because his father had been ordered to France to thwart the Red Menace. Nevertheless, an unlikely family friendship developed. "Once, we took him fishing. He really liked it; it wasn't tiring. We invited them to the grape harvest and to the harvest meal. They really enjoyed participating in those activities. Those are good memories," Boizeau said.

Interviews with Mayet, Gerbaud, and Boizeau throughout this book defined the big picture implicit in what they believe to be the Golden Ghetto legacy. But there were also snapshots supplied by individuals seeking one of the most fundamental of human needs—work. Two French women, one a teenager and the other a widow with a teenage son, defined the paradoxical allure of the base. The younger embraced all that was offered while the older, more worldly woman was forced to make sometimes distasteful accommodations. With base closure creeping ever closer, one faced the future with confidence and the other with uncertainty.

The younger woman was willingly seduced but realized that for the seduction to be complete she would have to offer enticements of her own. Learning English would be her ticket. The older woman, already in her mid-forties, disdained taking this step.

If you had imagined that only the most skilled and capable workers were to be recruited, you would be wrong. It would be hard to cite a better example than Paulette Prôt. She was among the thousands of French men and women who offered the Americans not even the barest workplace qualifications except for one—their willingness to work hard. Sometimes two jobs were needed if you were raising a family, but the jobs were there in a region where it seemed everyone was looking for work.

Paulette was ninety-six when I interviewed her at her small, impeccable, and beautifully furnished apartment in Touvent. She had lived there for almost a half-century, raising her son Jean-Claude, and was still going strong. For the interview, her hair had been freshly coiffed by her daughter-in-law Nicole and her tailored blue and white dress complete with a thin, black leather belt indicated how much she had looked forward to the interview. Wine was offered and it was not *vin ordinaire*.

"I came from Le Blanc and my husband had just died. I had a sister-in-law who lived in Châteauroux and I came to her house to find a job. My son Jean-Claude was thirteen. I rode in the school bus as a monitor, looking after the American students. The bus made two trips a day and I was home at two o'clock.

"After two o'clock I went and ironed and did housework for the first and only family for whom I worked. I can't remember their name. They had three children. I was always appreciated by everybody in the family. They were all very easy going. They let me go at six thirty. We were not friends. We talked to each other in the best way we could.

"I didn't buy anything American. Besides, I didn't feel like eating American food. You have to get used to everything but sometimes it's tough. When the family moved out, they gave me things they no longer needed. There were carpets and stuff like this, and I crossed Châteauroux with my boss in the convertible car! The family was moving to another town, Le Blanc.

"One day, with my son, we went to Le Blanc and we passed in front of her house so we stopped to pay her a visit. She was very happy so she prepared something to eat. She gave us some soup in which we found ham and prunes! During this time I had been working in a workshop where they made shirts in Châteauroux. I made mattresses. Then I worked in Le Poinçonnet where I made blouses.

"From the very beginning I always felt the Americans I met appreciated me. You know, you have to cope with any kind of situation. I was lucky to arrive in Châteauroux when the Americans were here because the jobs they gave me helped me very much when I brought up my son. The one big change after the Americans left, and you could see it everywhere, was unemployment. Jobs were hard to find and you were lucky when you found one."

Madelaine Dagot was eighteen and fresh out of school when she applied in 1952 for a job at La Martinerie. She had planned to work for only three months. It was fourteen years later and shortly before the base closed that she left. She was seventy-three when I interviewed her in her home in the former American enclave of Brassioux.

A petite woman, Madelaine, like other women I interviewed for *Golden Ghetto*, had obviously prepared for the occasion. Everything from her hair to her trim suit was in place. The American ranch-style house was neat and orderly despite the fact that there were overly curious grandchildren roaming about.

I told her what Maurice Garnier said about the salary cap at the base that limited pay to no more than ten percent of the wages paid in town. Madelaine laughed and waved away Garnier's explanation.

"We were well-paid and the salary had nothing to do with what people were paid on the economy. On top of that we had bonuses. If you spoke good English you had a monthly bonus. I worked at the PX. It was like a dream. I was walking on air. It seemed that I was like in a film, an American movie! My first job was in the food section and then at the cigarette counter. After that I was promoted to supervisor and took care of perfumes, jewelry, records, and photo developing."

For Madelaine, as with others, the ability to speak English paid big dividends over the years. It had transformed Madelaine into a woman of property. "I was able to put aside enough money to open up my own little household appliance shop downtown."

The work day for Paulette Prôt—who spoke no English or at best, very little—involved keeping order on buses carrying American kids to school, and washing and ironing clothes for a GI couple into the early evening. In her words, Madelaine Dagot, fluent in English, stepped onto a movie set each day for fourteen years. The work dividends were different for each of them. The payoff for Paulette included the thrill of riding across town in an American convertible car with the top down as it headed across Châteauroux to her home. The GI family for whom she worked was bringing to Paulette's home the carpet, household goods, and utensils they no longer needed and were leaving behind.

I discovered in writing this book that the unlikely became the normal, and as a result, the normal was hard to find. When fifty-year-old recollections surface, often under prodding, objectivity takes a beating.

I realized there were plausibility hurdles that had to be jumped because, as Pulitzer Prize-winning American novelist Richard Russo cautioned in a radio interview, "memory can be treacherous."

Memories can easily be in a state of freefall; anger could be elicited as easily as fondness. For some, their umbrage, if not total distaste, had not been lessened by the passing years. Pat Thacker, a CHAD High School student from 1961 to 1965, enjoyed four storybook years of endless activities and honors. The Free Masons were a very strong presence at CHAD and her father, Captain Jack Thacker, and mother, Mary Sue, were deeply involved in setting up the Eastern Star Ladies' Auxiliary Chapter at the base. Pat was elected Worthy Associate Advisor of the CHAD chapter of the Order of the Rainbow Girls in a ceremony that got front page coverage along with photos in the base newspaper. She wrote for the high school paper and her late brother, Edward, was one of the stars on the basketball team. It would be natural to assume that, given all of this, Pat would be a prime candidate to attend one, if not all, of the CHAD reunions that have been held over the years. Not so. Pat was sixty years old and living in Las Vegas, Nevada when I interviewed her in 2009. She explained why Châteauroux has never since been on her family's travel itinerary.

"My parents never went back to France or Châteauroux. They traveled a lot, but they never went back there," Thacker said. "There were two incidents that motivated their decision never to return. It was obscenities and so on. I was just a kid. My mother was spit on while walking downtown. It was by the Communists—they identified themselves as Communists. She was just walking on the street, shopping when it happened.

"She was furious. My dad was furious. You know my dad was a military man, and you know as a military man he did not like the Communists. His general reaction towards them was 'those damn Commies!' My mom and dad never gave me a clear answer as to why they didn't go back to France. But I think the spitting incident with my mother could have been the reason."

Sometimes it took much less than the time Mary Sue Thacker was spit upon by Communists to provoke a sense of anger. When pride was at stake, even the most mundane incident could do the job. It was a case of perception.

"I can tell you another story that probably contributed to my father's lack of desire to return to France," Thacker said. "After we moved to Châteauroux, our car followed us by several weeks. My dad led a group to Le Havre to pick up their cars, which had been shipped from the U.S. My father was a very smart man, went to college at fifteen, graduated first in his class from the Air Force Institute of Technology when he got his MBA, but he was not proficient in French. At one point, he was not sure of which road to take and stopped to ask a Frenchman which led to Poitiers. According to my dad, he gave five pronunciations of 'Poitiers' to this gent who mouthed back 'Poitiers?' Finally, the Frenchman said, 'Oh, Poitiers' and pointed them in the right direction. My dad had been pronouncing it fairly well, but knew the Frenchman was having a good time at his expense."

It might have been a minor, annoying episode, but it involved Thacker's injured pride and that made a big difference. As a navigator, he saw plenty of action in bombing raids during WWII. He transitioned from the old brown shoe Army Air Corps and was a longtime member of the privileged officer class, and as such, he demanded respect that an unwary and playful Frenchman had somehow violated.

For decades, reunions fostered the image of Châteauroux as a modern Shangri-La, forever young and vibrant. Two important reunions had already been held in Déols and Châteauroux, and a third was on the drawing board as this book was being written. Others had been held in Seattle, Las Vegas, San Antonio, and Steamboat Springs, Colorado. For various reasons, this back-slapping camaraderie was not for everybody. John Forman experienced some of it and did not want any more. As the teenage, bartending son of demanding and fun-loving General Red Forman and his wife, Boopy, John admitted he had enjoyed a privileged life during the family's five years at CHAD.

"I had been to two reunions and my wife told me, 'you spend too much time in the past.' She's probably right, it was a lot of fun, but that was a long time ago. As much as I enjoyed it, why would I want to go back, what would I go back for? I think that it would have ruined the whole image I had of those years. I wanted to keep the dream alive," John said.

Thinly veiled resentment was easily discernable when Janine Egtvedt described what had happened to Châteauroux after the forced abandonment of CHAD. Golden memories, the happiest of her life, had been violated. Janine, a gendarme's daughter, met and fell in love with Jim Hawkins, the son of a Kentucky tobacco sharecropper. Despite their divorce, Janine said that Jim remained the love of her life. Much like Baudelaire's swan, she discovered on her first and only return visit to Châteauroux that all of the couple's old haunts had disappeared forever. "I saw that the bars were closed and that was important because that's where the money was. And second in importance were the restaurants and also the shops. GIs and their families would go to all of these places and spend money. These were the places that Jim and I loved to go to. When I went back, they were all gone. You didn't see very much activity. It was sad to see. I saw no reason to go back there again. I don't want to go back," said Janine, a Châteauroux native whose family still lives in Berry.

The Dreaded Day Arrives

Leonard "Lenny" Kalina arrived at CHAD in 1960, nine years after the base had opened and at a time when it was easy to discern the economic evolution that was well under way. During his six years at the base, he witnessed how the "amazing number of people on bicycles pedaling back and forth" became fewer and fewer, giving way to the ubiquitous Deux Chevaux and hundreds of cheaply priced Renault Dauphine. Although pedal-power had decreased in scope, it was still important—but Kalina was not. De Gaulle's proclamation made him superfluous. Kalina ran the highly classified, seldom talked about International Telephone & Telegraph (ITT) operation at CHAD. Its purpose was to make sure that hi-tech communication between the U.S. Air Force bases scattered over three continents was flawless. The fifteen crews of two to four men each that Kalina dispatched to do the job were among the first to pack their bags and disappear from CHAD. The guys that Kalina had working for him might have drifted away, but it was not so easy to disburse the millions of dollars in hi-tech equipment he had on his hands.

"When I got all our stuff together for shipment, the base was almost deserted. It was down to a skeleton force," Kalina said. "We had to evacuate the base and find a place to ship everything. I was responsible for packing equipment for shipment. Everything went to the Strategic Air Command (SAC) base at Zaragoza, Spain. I was one of the last civilian contractors on the base when it closed."

While Kalina and a handful of his highly paid civilian technicians were nailing shut large wooden shipping crates, students at CHAD High School were told that the 1966 graduating class would be the last. Some desks were left behind, but basically the school was stripped. The Beaver Cleaver and Ozzie Nelson community of Brassioux was methodically abandoned. Families were given their travel orders, told to pack up their essential personal belongings, and leave behind what was not needed. It was the same at 410 housing, whose vacant apartments, built to French standards, would go a long way to ease Châteauroux's housing shortage. Non-Berrichon saloon and restaurant operators, who had followed the money to Berry, were among the first to close up shop and head home to Paris, Toulouse, or Tours. All but a few civilian contract workers were gone, and as Thomas Ciborski recalls, members of the rapidly dwindling military contingent were marginalized by their bosses at SHAPE (Supreme Headquarters Allied Powers Europe), if indeed they were given any thought at all.

"We ate a lot of canned ham—ham for breakfast, ham for lunch, ham for dinner," said Ciborski, a business administration graduate of St. Francis College in Brooklyn. "We had plenty to drink, vodka, beer, and gin. Especially beer, because the guys we sent to get food rations bought cases of beer instead. The trucks making the deliveries were driven by the Army, a bunch of kids eighteen to nineteen years old. There was a lot of drinking going on, and these kids were driving these big trucks that we'd sometimes find beside the road in a ditch. Everything was in a state of flux, everything was chaotic.

"There were only about a hundred people on the base. There were people coming and going all the time. These were the people, Army and Air Force guys, who were actually closing down the base, packing up crates, taking heavy equipment such as cranes and transporting them to other bases. They were taking almost all of the military vehicles and getting around the base was tough, unless you had your own car.

"It was really laid back. We had no formations and were billeted in the Bachelors Officers' Quarters where we each had one room to ourselves. We knocked off at four o'clock. I really don't know who was in charge; it could have been a master sergeant for all we knew.

"As you would expect, things could get wild with a hundred young guys with a lot of time on their hands and no real supervision. One night we pulled a pay telephone off the wall outside a café in downtown Châteauroux. We had already been spotted by the French police when we jumped aboard a French bus with the telephone and headed back to the base. The gendarmes were right behind us, but the bus driver couldn't care less. He got us back to the base okay."

Ciborski, stationed at CHAD from September 1965 until just before the base closed, was cashier during the final months when he made a discovery that illustrated how the base closing could take on the aspects of a Laurel and Hardy comedy routine.

"Because the base was closing down, American currency was being packed up and shipped out, except for one wooden box with between five thousand to ten thousand dollars," Ciborski said. "We notified the proper people, and to our surprise, they flew in a C-130, which is a really big plane. We went out in the rain to meet it, packing .45 caliber pistols under our raincoats, hoping that we wouldn't shoot ourselves. An officious, self-important officer, I believe he was a major, got out. He was packing a .45 caliber pistol. The freight door opened and a squad of heavily armed MPs piled out and set up a protective perimeter. They were even wearing camouflage.

"I couldn't believe all of this was for one box containing maybe ten thousand dollars. Here we were in the middle of the tarmac with that one wooden box of money on the ground. When the major saw what we had, he really got pissed off, calling us idiots. It took them probably about five minutes to load that one box onto the plane and fly back to their home base in England. They had misunderstood how much money we had, believing that there were a dozen boxes with ten thousand dollars in each one. The money they got wouldn't cover the fuel cost of their flight."

Transfers had decimated Ciborski's fun-loving group, and by Christmas 1966, there were twenty of them remaining to chip off the rust, repaint, and refurbish a playground swing set that had been left behind. The swing set was to be the nexus of a Christmas visit to a Châteauroux orphanage where the airmen, well fortified with Christmas cheer, were to present individual presents and a donation of four thousand dollars that had been collected at the base.

"There were a lot of young kids between, I'd say, six to thirteen years old. The kids were quite guarded, maybe even a little scared of us. In our group there were twenty young guys who had been drinking all morning, so it was easy to understand. We even brought along our own Santa Claus," Ciborski said. "With all of our good intentions, this could have fallen flat. When we gave them each a present, they kind of stared at it like they didn't know what to do. There was an awful lot going on. This was all new to them. Some of them went out to the swing set, it was very cold. Their clothes looked worn, like hand-me-downs. The orphanage put on a really good spread for us, good wine and fine French food. A few days later, we got a letter of appreciation from the l'Indre Director of Medicine and Social Services thanking us for the 'kindness and generosity that enabled the children to have a warm and wonderful holiday.'"A letter such as that from a regional bureaucrat did little to assuage a feeling of resentment that had been growing among the American military ever since Charles de Gaulle placed sixteen years of NATO safety guarantees in a time capsule and buried them. This resentment applied not only to desk-bound Pentagon brass, but also to the boots on the ground at CHAD. Some sort of "payback" was necessary, but how and what? Symbolic gestures would not do, and as a result two precipitous things took place. Mike Gagné explained:

"There used to be a railroad with tracks going over this little bridge across the Highway, D-25. Just past this bridge on the left, there was this huge quarry where the Americans buried all kinds of equipment, electrical and other stuff from the base. It was pretty big, I would say it was between 250–300 feet wide and 50–60 feet deep. I remember that they used to take huge limestone blocks out of it, and I believe it was owned locally. Basically, it was everything they didn't want to bring with them when they left. They dumped it into the quarry, burned it, and then used bulldozers to completely cover it with dirt.

"If you're looking for an explanation why they didn't give the surplus stuff to the French, I think it's simple. Hey, the Americans got kicked out and of course they didn't like it. I was told that they even took all of the light bulbs with them. I don't think they would have left a jeep."

A secretary at La Martinerie during the final months, Madame Lucie Pee, eighty-three when interviewed in 2009, was forthright. Americans arrived in France from what she believed was a land of plenty, therefore, they could be judged as much for what they did as well as what they did not do. When asked about the burial pit, she said, "There were even fridges buried. They left in a few days. It was very fast. When General de Gaulle said it was over, they started the removal. I don't know if it was reprisals. It was very shocking for us. One day I left work, and the morning after, some buildings had been destroyed in the night."

During the interview, Madame Pee, who was obviously exacting a little "payback" of her own said, "They had everything at the base except a swimming pool which was built by the French after they left. The saying goes that the white Americans were not in favor of a swimming pool to avoid swimming in the same water as blacks."

Gene Dellinger, who was among the first GIs to arrive at CHAD in 1951, found himself back at the base in 1967, an assignment that he had requested and had surprisingly been granted. Dellinger and his wife, Monique, had strong attachments to the base, and he did not feel that it was macabre for him to be there when the Stars and Stripes came down for the last time. While airmen and their families were being shipped out in large bunches, Dellinger stayed behind as he had requested, as part of a small contingent under the command of Colonel James Fahey, a specialist in Air Force base closings. "Colonel Fahey was the last base commander. Whoever selected him to close down the base couldn't have done better, he was the perfect man for the job," Dellinger said.

The base was being dismantled piece by piece, and although the buildings would remain, whenever possible, nothing of value was left behind in their carcasses. Dellinger saw it all.

"Not all of the stuff was shipped out. I knew about the burning pit, but I never went out there because I didn't want to see what they were doing. But I did know what they were doing. I did see the smoke coming from the pit where they were destroying equipment. It had to be destroyed and couldn't be left behind as trash. They would burn the stuff in the pit and then bulldozers were used to cover it over." Frustration surfaced whenever dismantling efforts were thwarted, as was the case at the former base hospital.

"Congressman Ralph J. Rivers, a democrat from Alaska, burst on the scene to find out what was happening to all of the material," Dellinger said. "He jumped off the train when it arrived and literally ran across the tracks at the station to get to the base. I'll never forget, he went into the former hospital building and he saw all of the furnaces and radiators. He said: 'We can't leave all this stuff here, it's got to be taken out. The entire heating system has to be taken out. It's still good and it has to be removed.' Of course, it didn't come out. Colonel Fahey discovered it would cost too much to remove and ship it, so it stayed."

The air of reprisal that permeated CHAD was not an isolated phenomenon. Canadian airmen also found it hard to accept the de Gaulle decree after traveling thousands of miles to help defend France for the third time in the last half-century. The Royal Canadian Air Force (RCAF) had taken its blinders off long before the Americans, and in Fall 1963, they were already preparing for the inevitable. Bob Hallowell, in charge of mission planning and intelligence at Grostenquin Air Base, recalled, "After Big Charlie [de Gaulle] said 'no Yankee nukes' [in France], I knew our days were numbered. Two weeks later we got the word: 'everybody out.'"

French Army Major Pascal Enon was asked his opinion as to whether the burning and burial of American equipment could be considered an act of reprisal. "Reprisal? Of course it was, if it actually happened.

This is exactly what I think," said Enon. Surprisingly, and without being prompted, he described what he had found when he was stationed at a former RCAF base in Alsace near the German border. "When the Canadians left many years before I arrived, they poured some concrete into the boilers. They even did that into the toilets. I did see the result. It was obviously reprisals. They didn't want the French to use what was left."

STEINWAYS, TRACTORS, AND NO MORE FREE COFFEE

Dellinger was part of a crew whose job it was to see that very little, if anything, was left for the French. He was amazed at how much had to be done, the vast amount of material and equipment to be cataloged and shipped out, and how little time remained to get it all done.

"First, we had to discharge the final French workers, possibly two hundred or so. Many of them wanted to get out quickly so they could find a civilian job," Dellinger said. "But before that, there were huge warehouses that had to be emptied. Everything in them was listed by category. We found big, concert Steinway pianos that had never been used, and tractors that were to be sent throughout Europe that were never uncrated. Planes would come in and Fahey would say, 'This load goes to Spain, this load goes to Germany,' which got the best stuff we had. England got the next best, and the worst went to Wheelus Air Base outside Tripoli in Libya. Would you believe it, when Châteauroux closed, Fahey's next command was Wheelus Air Base, where he had been shipping the worst stuff."

Silent embarrassment accompanied the systematic break-up and eventual abandonment of CHAD. It seemed that no one wanted to openly discuss what was happening as building after building was gutted and some structures were destroyed. It was a thankless job. Dellinger recalled waking up one morning and discovering that Fahey had left during the night without a word or hint to anyone that he had packed his bags and was about to disappear. "I went to the base engineering building where Fahey had his office, and discovered that the only man there was an old, brown shoed staff sergeant," Dellinger said. "I told him I worked for Fahey and that he had left. As a result, I had nothing to do, so the sergeant put me in charge of base security.

There were only a few of us, but it gave me something to do. We had a deal where we ate our meals at Joe's From Maine in Châteauroux and we had our office set up in the base fire station. It was basically a ghost town.

"In the final days, the only thing that was left was the golf course, and the few GIs who were left would go into the clubhouse for coffee, snacks, and sandwiches," Dellinger said. "This was at the pro-shop that by this time was being run by the French. It had all been prearranged earlier. One morning, one of the guys came up to me and said, 'They won't give me any coffee.' So I went over to the pro-shop with him and there was a French employee behind the bar who said, 'We're not serving GIs coffee anymore. You're not even authorized to be in here.' It was the first that I learned it was a new French restaurant; that it didn't belong to us anymore. She said, 'I've got orders not to give the GIs any coffee.' I told her, 'Well okay, if we can't get coffee and we're not authorized to be here, then you're not coming in because I'm locking the gate.' From that point on no one from Châteauroux would be able to come in to play golf."

Paradox and irony defined the shift in power at CHAD during its closing months. It might have been only a grizzled, former brown-shoed staff sergeant who gave Dellinger and his three-man crew the seemingly menial task of patrolling deserted streets, empty offices, piles of debris, a hospital without beds, and warehouses echoing their emptiness through yawning doors. But with the job came the ultimate symbol of authority: keys to the gate. Once the French golf course operators realized this, "Things were corrected very quickly. We got all the coffee we wanted."

For sixteen years, it would have been hard to find a more successful example of Franco-American Cold War bonding than CHAD. Certainly, there had been political and philosophical differences and national pride could take a beating. Thousands of Americans from lowly GIs to two-star generals, civilian employees, and contractors, along with their families, passed through its gates. At one time, one out of every six French working adults in Berry picked up a Golden Ghetto paycheck. During its death rattle it had come to this, a dispute over a handful of GIs being denied a few cups of coffee.

Conflicting statements confirmed that just as its birth pains were hardly pleasant, the golf course could not go gentle into the good French night without controversy.

Former airman René Coté was selling and renting a wide range of cars while pedaling auto and personal insurance in 1966. "They simply rolled the putting greens up. That's right, just rolled them up and loaded them onto flatbed trucks. Then the Red Ball Express drove them off to Germany to be used at another Air Force golf course. And the clubhouse, which was supposed to become a French version of a country club—of course that never happened either."

Bernard Maillard, who in 1967 was a lieutenant at Châteauroux, serving under Colonel Claude Delbecq, and would eventually become base commander, remembers it this way: "The greens were removed by the Americans and stored somewhere in town. They were supposed to be shipped to Germany, but there were negotiations and after a while it was too late. Everything was rotten. Before everything was removed, we asked for a member card [from the French operators] to be able to practice golf, but the price of the card corresponded to one month's salary, so we never played golf. When everything was removed, we transformed the place into a maneuver ground. It was not possible to imagine people playing golf near soldiers shooting."

It was still being debated decades after the closure of the golf course whether the "shooting" referred to by Maillard would have ever occurred if there had not been a falling out between the civilian golf course operators and the French military. "The difficulties with the French military were apparently serious," said Dellinger. "The base commander apparently had enough. He realized that the golf course fairways were a great place for maneuvers and that's what he did. He had his regiment, I believe it was an infantry regiment, with all of its trucks and support machinery conducting maneuvers right across the golf course—tore it up. And that was the end of the golf course."

Forty-two years later, there is no evidence that a golf course ever existed at the base. What had been a barren, weed covered field when

General Smith first envisioned broad sweeping fairways, well-tended sand traps, excellent putting greens, and soaring golf balls, was once again an overgrown field. It is doubtful that after a few more years there will be anyone still alive who remembers that a golf course had even been there.

During the 2007 reunion, a sizeable group of returning pilgrims drove past the large, non-descript field that had once been the golf course without so much as a passing glance. Theirs was a single-minded mission: to visit the dilapidated CHAD High School buildings, enshrined ruins that illustrated the hold that the Golden Ghetto still exerted four decades after its closure. Mike Gagné escorted a group that included former students Janice and Sheila Witherington, and former high school principal Cliff Gunderson and his wife. "We went into a classroom where a lot of the floor tiles were broken. Big and small pieces had separated from the floor, and members of the group bent down and started collecting them," Gagné said. "That's right, they were collecting shards of tile more than a half-century old to bring home as souvenirs. At first I thought they were nuts, but then I got caught up in the whole thing and said, what the hell, and started picking some up for myself."

Gagné had never found Châteauroux to be exceptional in any way, and often wondered aloud what had turned a long ago Golden Ghetto into an ageless tribute to youthful dreams. Dellinger, on the other hand, found it hard to stay away, serving four tours of duty that began with him and his companions sharing a tent and a potbellied stove during the winter of 1951–52, and then finally guarding a relic.

After his tours of duty at CHAD, years that he described as the best of his military service, Dellinger found himself retired, out of work, and looking for a job. He filled out a three-page State Department application after hearing that a caretaker was needed for the American cemeteries along the Meuse in northern France. "I filled out the application and kind of forgot about it," said Dellinger. "I figured there would have been hundreds of applicants, and I was really surprised when they notified me that I had been hired." When I interviewed him in 2009, he had retired after fifteen years as the

American caretaker of the famous American gravesites at Normandy. He hosted every American President from Jimmy Carter to Barack Obama. But most of all, he remembered the day the Stars and Stripes were lowered for the final time at CHAD.

"The flag came down for the final time on March 23 [1967]. After we took it down, we gave it to a museum, and today that same flag is at the Peace Museum at Verdun, which is a good thing because it will be there forever."

Apparently, the flag never reached its destination, and why it never got to the *Mémorial de Verdun* is an example of how capricious historical predictions can be.

The only thing that is known for certain is that the flag was handed to Michel Aurillac, who then gave it to Charles-Armand de Gontaut-Biron, who was in charge of *Le Musée des Trois Guerres* in Diors.

When the Diors museum closed, its indoor collections, which included the CHAD flag, were sent to *Conseil Général de la Meuse*, the county along the German border where the Memorial of Verdun is located.

Yves Bardet, the Brassioux historian, contacted the *Conseil Général de la Meuse* and was told that the flag was not there.

Isabelle Remy, who is responsible for the Center for Documentation for the Memorial of Verdun, said no documentary evidence could be found that the CHAD flag had ever arrived.

So the conclusion reached by the *Conseil Général de la Meuse* was that the flag had never left Diors. If it had indeed left, it had never arrived at the Memorial of Verdun. If it did arrive, the flag was probably packed away to gather dust along with other forgotten military relics.

1966 – COLD WAR POTPOURRI – 1967

- On January 1st, 1966, despite strong tobacco industry opposition, the U.S. Surgeon General decreed all cigarette packs must carry the warning "Caution: Cigarette smoking may be hazardous to your health."

- On March 4th, a London paper quoted John Lennon as saying ". . . the Beatles are more popular than Jesus," prompting worldwide hostility toward the group that led many people to destroy Beatle records.

- On May 16th, the Great Proletarian Cultural Revolution began in China, eventually reaching America, where Mao's "Little Red Book" became required hip pocket reading for angry university students.

- On November 14th, the name Cassius Clay became history when sportscaster Howard Cosell honored Mohammad Ali's wishes to be called by his Islamic name during a post fight interview.

- On January 15th, 1967, The Rolling Stones made their first appearance on *The Ed Sullivan Show* but only after agreeing to change the lyrics of "Let's Spend the Night Together" to "Let's Spend Some Time Together."

- On January 27th, 1967, while on the launching pad at Cape Canaveral, Astronauts Virgil "Gus" Grissom, Ed White and Roger Chaffee died in a flash fire aboard the *Apollo I* spacecraft.

- On July 7th, to counter mounting anti-war protests, General William Westmoreland, U.S. military boss in Vietnam, said "North Vietnam is paying a tremendous price with nothing to show for it."

- On December 3rd, Dr. Christiaan Barnard performs the first heart transplant in Cape Town, South Africa on Lewis Washkansky, 53, who died eighteen days later of pneumonia after his immune system failed.

EPILOGUE

Châteauroux's weather on the early morning of June 20, 2010 had everyone worried. There had been intermittent rain and plenty of thunder and lightning during most of the preceding three weeks. The skies were overcast, not a good sign if you were planning an outdoor ceremony.

Throughout the morning, onlookers had begun to assemble on the sidewalks around a grass-covered traffic circle in the middle of which stood a ten-foot tall, cloth-covered object, which had been anchored to a cement foundation. A temporary wooden walkway extended across the grass from the surrounding street. Across from the grassy knoll, microphones and amplification equipment were standing at the ready. It promised to be a warm day. A café two hundred yards away was crowded with coffee and apéritif drinkers awaiting the ten o'clock unveiling of *La Flamme de L'Amitié* (The Flame of Friendship).

Among the arrivals were over sixty American pilgrims for whom this ceremony would be the culmination of an odyssey that for many began at the Châteauroux Air Station High School. The French Army would soon be pulling out of La Martinerie once and for all, ending a military presence that dated back to 1917 and included not only the French, but German and American forces as well. Their guided tour of CHAD two days earlier would be their last. Plans for the event began in August 2008, when an anonymous former CHAD High School student approached Janelle Peterson-Wolford, who was President of the Châteauroux American High School Alumni Association and the creator and webmaster of the two most prodigious CHAD websites, one devoted to the Air Station and one devoted to the high school.

Cheers and clapping accompanied the unveiling. The modernistic bronze sculpture depicts two intertwining flames reaching skyward. To either side, along with others, were the two sculptors, David Newton and Tomas Bustos. The Flame of Friendship represented the first major international cooperative commission awarded to two American minority artists.

The studios of Bustos and Newton are within walking distance of each other in Dallas. Newton, a black graduate of the College of Creative Studies in his native Detroit, studied in Italy and roamed the famous galleries of Europe. He moved to Dallas fifteen years earlier after winning a national sculpture competition. Bustos's artistic abilities caught the attention of his high school teacher in Dallas. Acting on the teacher's recommendation, the renowned sculptor and teacher Octavio Medellin jump-started the eighteen-year-old Bustos's career when he took him on as an apprentice in his studio. Newton and Bustos began their collaborative efforts in 2001. On an overcast morning in central France, almost a decade of effort was being rewarded with an important international commission.

"This event has been in the making since August 2008," Janelle Peterson-Wolford told the crowd. "Our benefactor, the anonymous donor, asked me to organize this event and make it happen. His fondness for the Berry Region is deep and this is his way of honoring his youth and the great times he experienced here."

Making it happen required more than a year of negotiations. On the surface, it would seem that presenting a gift would be a matter of simply accepting it, expressing appreciation and, in the case of The Flame of Friendship, prominently display it for posterity. But in France, accepting a gift valued by the newspaper La Nouvelle République at 150,000 euros, had to satisfy French bureaucracy's belief that even a free ride necessitated paperwork, and a lot of it. Finally put to rest were questions involving import duties, taxes, permits, jurisdictional and municipal boundaries, and the go-ahead was given for the Bustos-Newton creation to be put in place.

Bustos and Newton felt honored to accept their commission, but until their arrival in Châteauroux they were somewhat bewildered about why an anonymous donor would pledge this kind of money for a bronze sculpture to grace a traffic circle in an obscure French town.

"It was a mystery to me why, after all this time, they still had such a strong emotional attachment," Newton said. "But last night I truly began to understand it. We were welcomed into the home of a well-known regional artist who signed his works, L'ours. He showed us his gallery, and being sort of a wine connoisseur, he really rolled out a good wine. We had *pâté de foie gras*, he did all the cooking himself and the baking. He baked pastries, bread, and opened up his entire studio to us, and we were strangers. So I said to myself, 'Okay, now I'm beginning to get it.' This is probably an example of what the people of Châteauroux did back when the Americans were here."

"I knew absolutely nothing about Châteauroux. I picked up a phone one day and Janelle asked me if I would be interested in doing a sculpture in France," Bustos recalled. "After seeing examples of my work on the Internet, she came to my studio and saw the project that David and I were working on and she liked it. So that's when we came up with ideas. And that's how it all started."

After having his first two proposals rejected by Janelle and the donor, Bustos came up with the idea for the friendship flame. "When I showed her the drawing, Janelle's eyes lit up. So I did a three dimensional mock-up that was about twelve inches tall, and she loved it. I knew it was a 'go' from there.

"But you still don't know how it's going to be accepted until the unveiling. What I tried to impart was to give the people of Châteauroux a sense of what I feel, a sense of movement," explained Bustos. "Kind of what happened here, the warmth of people working in friendship to build an airport. The people we've met have been very, very warm and friendly. And I really enjoy working with them in putting this whole thing together, all of it with a feeling of warmth and friendship."

Newton described how much The Flame of Friendship meant to his and Bustos's careers. "It is the first international commission either of us has ever received, and it is in Europe, so that's a big deal. So now you can call yourself an international artist and you're telling the truth. Remember, you only need one big commission to say that."

International solidarity dominated the dedication ceremonies. French and Americans stood in silence while their respective anthems blared from portable speakers and the Stars and Stripes and French Tricolor ruffled in the late morning breeze. There was a surprisingly short list of speakers, including the mayors of Déols and Châteauroux, Carly Van Orman, representative from the U.S. Embassy in Paris, and a big thank you from Janelle Peterson-Wolford. French Senator Jean-François Mayet recalled his childhood bonding with American kids during the mid-1950s.

"I had realized that it had been only ten years since fifty thousand Americans, maybe your parents or your friends, had lost their lives on our French shores to liberate us from foreign occupation which we, the French, had not handled in an exemplary way," Mayet said.

This comment set off a reactionary buzz among more than a few French and American listeners. During my five years of interviews and research for *Golden Ghetto*, it became obvious that any discussion of how the Vichy government and French civilians handled themselves while under German occupation was taboo. The several times that I broached the subject I was politely cut off. The fascist, pro-Nazi Milice was an easy target, but it never went any further than that. Mayet had cracked a seldom opened door and his words bounced from table to table during the elaborate luncheon that followed. The senator was equally blunt in describing how de Gaulle's NATO eviction notice affected him and his youthful friends.

"And then everything came to an end. We were left with only memories of fabulous moments made of much friendship, flirts, and sometimes love. A sad and silent veil dropped over the town. Goodbye, beautiful cars, planes, and above all, jobs. Some French people—very few—left to live in the U.S.A. Others, like myself, had dreamt of doing so but had to satisfy themselves with waiting for a surprise visit from some American friends when they came to Europe for a visit."

Barring either a man-made or natural catastrophe, the Flame of Friendship should be sitting on its traffic circle pedestal forever. It is doubtful that this sculpture will meet the same fate that has befallen the Stars and Stripes *Forever Flag* that was lowered for the final time March 23, 1967. Then Master Sergeant Gene Dellinger predicted the flag would become a permanent part of the peace memorial at Verdun.

With the fate of the *Forever Flag* a mystery, the mission of the Flame of Friendship has taken on additional meaning. Although it can never be unfurled for parade and ceremonial duty, it will be around for a long time to remind passers-by that Berry was graced by a Golden Ghetto for sixteen years.

SPOUSE SEEKERS AND JOBS

The first chapter of this book included two personal ads that illustrated how Berry's economic good times were in full swing during the early 1950s. Economic prosperity, fostered by the Marshall Plan and protected by Truman Doctrine military muscle, provided clear guidelines for young, husband-seeking maidens. One young lady cautioned that men need not reply unless they could produce a photo of their tractor.

An affluent farmer boasted about the size of his property and that his twenty-two year old daughter was available for marriage, along with a sizeable dowry; don't bother to reply unless you are serious, with a clear implication of what serious meant.

Berry, along with the rest of France, had enjoyed decades without a war, an economy that progressed nicely from year to year and, with socialist assurances, long vacations, and a bundle of holidays, people were living longer and marriage guidelines had changed. Certainly, there were plenty of young, spouse-seeking men and women, but in 2010, the geriatric set was taking its desires public.

Sober farmer, non smoker, quiet, serious, would like to meet a young woman, 45–55 years, who is celibate, lively, divorced, honest, serious, sensual and kind.

Widow, soft, pleasant, elegant, dynamic, good health, rural owner, looks for friend, non smoker, sober, cultivated, 78 years maximum, to break loneliness.

Thousands of jobs for French workers bolstered a regional economy that grew more than 33 percent in a few years, thanks to Uncle Sam. When the base closed, the jobs disappeared, the middle class shrank, and the economy plummeted.

These factors were very much in evidence when it was decided to revisit students studying for their baccalaureate at Châteauroux's most prestigious public *lycée*. Four years earlier, top students at the same school were asked what they knew and understood about the American presence in Berry. Their answers during the earlier interviews were based on what they were told or had read, and in 2010 their answers were based on what they were seeing around them. The question was whether they would remain in Berry when their school years were over. They responded only after it was agreed that they would remain anonymous. Their answers were depressingly similar, and the following examples were typical.

"After my studies, I don't think that I will come back to Châteauroux because I want to work in a big firm or in a college in a big town. In addition, I find this town very old, there are not lots of young people. Châteauroux isn't very dynamic."

"I don't like Châteauroux very much. There aren't lots of activities, while there are more attractive towns or cities not far away from here, like Tours where I will study next year. But who knows, maybe I will miss Châteauroux one day."

"I don't think I will work in Châteauroux. I prefer going to the countryside around Châteauroux, but not downtown. My wish is to leave the region so I don't think I will stay here."

"My parents live in Châteauroux, but I don't want to live here, I just want to come back and see them."

These interviews were conducted a few weeks before what could prove to be a significant economic rescue mission for Berry was debated. It seemed that it was more than a wishful rumor that the Chinese were considering Châteauroux as another way-station in the growing globalization of China's economy.

THE CHINESE ARE COMING . . . PERHAPS

At the time this was being written, there were many rumors. Would there be ten commercial transplants from China, or would it be thirteen? Nobody really knew, but it all made for enthusiastic speculation. Regardless, a Chinese manufacturing or commercial presence in Berry was viewed by many as a pot of gold at the end of the region's diminished economic rainbow, but not by all.

The stories about the Chinese are "bullshit," said Louis Morin, the self-proclaimed bourgeois businessman from Levroux. "I've talked with Chamber of Commerce associates in Châteauroux and that's what a lot of them think."

It was Morin's contention that the Chinese would be bringing pre-packaged kits from China for "do-it-yourself furniture" to be assembled by French consumers. "The 'Made in China' label has become a joke, all they want is to be able to put a 'Made in France' *marque* on the stuff they are shipping over."

On November 27, 2009 the fact that China's Communist-Capitalists had their eyes on Châteauroux became a *fait accompli* when Châteauroux and several neighboring towns signed cooperation agreements with potential Chinese investors. The company that headed the list of Communist entrepreneurs is Jimei, a furniture company established in 2005. Yujie Liu, Jimei's Vice President General Manager, defined her company's immediate plans, "We would like to come here to create our European platform to promote Chinese up-market products, and thanks to this platform we will also be able to import French products to China."

Senator Jean-François Mayet can count several travel notches to China on his French passport, each trip part of a continuing effort to bring Chinese business to Berry. The missions seem to be bearing fruit. "In Gran Déols, Jimei intends to develop a showroom to adapt Chinese products to the European market. We can expect the creation of about 400 jobs with at least 80 percent of French workers," Mayet said.

If we are to acknowledge centuries of business acumen, the Chinese do not believe in advancing with only small steps. "We also would like to build houses near the site to house our employees and their families if we have the possibility," said Mme. Yujie Liu. It may be no coincidence that the Jimei conglomerate includes a travel business with golf resorts and cruise ships, and that Chinese tourists have become ubiquitous and France has become a favorite travel destination.

Martin Fraissignes, general manager of sales and marketing for the Dassault Airport, had also been collecting big bunches of frequent flier miles with his trips back and forth to China. During these trips, Fraissignes also gave classes extolling Berry's economic potential. For years, he had been trying to drum up worldwide air cargo business for the former U.S. Air Force base. Its runways had been lengthened to accommodate even the largest passenger and cargo planes. This includes the Airbus A380, which, with a push here and a shove there, can pack in 850 passengers. A more humane version, Airbus A380 Model 007, is configured for 520 passengers. Fraissignes undoubtedly would love to have the refrain "the Chinese are coming," become a reality.

"We have what is needed to make this a reality. If the Chinese are coming, it makes sense that they would fly to Déols," Fraissignes said. "There are other possibilities, of course, but why would they offload elsewhere and then transport their product overland to Châteauroux. I don't think that would be sensible."

Fraissignes is only the latest to attempt to make sense of what to do with the old Air Force base. Back in 1966, only months before the Châteauroux-Déols U.S. Air Station was to become history, gruff-talking, hard-drinking base commander Colonel Francis J. Pope did not attempt to hide his pique that a decade and a half of American commitment was needlessly coming to an end. Pope was not a transportation and procurement kind of guy, but a former fighter pilot who believed that being blunt left no room for misunderstanding. "We are leaving you the best and biggest airport in Europe. It has everything. It's centrally located, all of France is only four hours away at the most. Now do something with it!"

Nick Loverich recalled that the local community and business leaders attending the luncheon responded to Pope's admonition with deafening silence.

If the Chinese were indeed on the way, then a half-century of wishful thinking and hand-wringing might be coming to an end if only in a small way. The Jimei furniture assembly facility might employ only 400 workers, but Mayet and Fraissignes believe it could be the start of a much longer economic wish list. It might never be the same as the glory days of the Golden Ghetto, when one out of every six of the region's French workers were picking up paychecks from Uncle Sam.

A reader might ask, what does the anticipated arrival of the Chinese have to do with the forced abandonment of an American air base almost a half-century earlier? Well, for those with a sense of history, plenty. It was noted at the very outset that this book was replete with irony. When General Joseph H. Hicks and his small entourage set up shop at the St. Catherine's Hotel in February 1951, it was less than seven years since thousands of American GIs had died on the Normandy beaches, and thousands more lost their lives driving the Nazis from France. Twenty-seven years earlier, more than 320,000 young Americans were killed and wounded in the bloody trenches of the Western Front while helping thwart Kaiser Wilhelm's assault on French freedom. In 1966, Charles de Gaulle decided that all of this carnage notwithstanding, the Americans, along with all other NATO forces, had to go. So he gave them the boot.

History is capricious. Hicks's arrival in Châteauroux was a little more than three years before the French defeat at Dien Bien Phu. The loss provided the death sentence for France's Southeast Asian empire. The victory by the Vietminh under General Vo Nguyen Giap was made possible to a very large extent by assistance from the Chinese Communists.

In 2010, the irony implicit in the hoped-for Chinese economic revival of Berry was evident. Communist entrepreneurs had taken options on several sites and it looked as though a furniture factory was on its way. Fifty-six years earlier, Chinese Communist military bosses decided the French had to get out of Indochina and took steps to see that it happened. It helped that French military commander General Henri Navarre and his artillery commander at Dien Bien Phu grossly underestimated the artillery capability of Giap's forces.

They had not reckoned on Chinese assistance to the Vietminh. Twenty 105 millimeter howitzers captured from U.S. forces in Korea were hauled through one hundred kilometers of jungle and placed strategically around the French forces. The Chinese sent along instructors to teach the Vietminh how to use them. At the height of the battle, Giap had 288 guns and mortars, the French had 88. In 1954, Mao Tse-Tung's military supplied heavy artillery needed to drive the French out of Indochina, and in 2010 President Hu Jintao's government-approved Capitalists were being wooed by the French because of their new-found economic muscle. It would be hard to argue that the transition from heavy artillery shells to furniture was not a good one.

In Spring 2010, linkage for the Chinese Capitalist connection was yet to be seen in central France. If and when it is completed, how enduring will it be and what place will it hold in Berrichon memories? Would there be a fond comparison with the sixteen-year American presence during the height of the Cold War? Almost a half-century after the Americans left, the French who had benefitted from what Uncle Sam had brought to Berry, had a hard time letting go of the Golden Ghetto which had been created in their midst.

Claudette Chardonner Neveu, who worked as a teenage nanny at the base, was one of those who refused to let the CHAD heritage fade. She married an airman, raised a family while living in Ohio and Iowa, and was divorced after twenty years, returning to her native Châteauroux. Mme. Neveu confided during the Flame of Friendship ceremony that she belonged to *Les Anciens Employés Civils de La Martinerie-Déols*. The group's two hundred members joined together in January 2008, motivated by the conviction that, during their lives, they would not allow CHAD to disappear as a historical footnote. "Yes, you could say that the members don't want to let go of what they once had," Claudette said. "That's what brings us all together."

THE VERY, VERY LAST PILGRIMAGE

It was a reluctant bunch that got out of the big tour bus at La Martinerie on June 18, 2010. It was evident that, although unspoken, Shangri-La was about to come to an end for them that morning. This stop was in a parking area that fronted a row of the largest warehouses they were likely to ever encounter. The vast doors were open and although there was only a light breeze, it was enough to send a hollow, somewhat eerie moan through the buildings.

Former enlisted airman Richard Shilling was among the pilgrims who gaped in awestruck wonder at the mammoth storage buildings. Shilling's three-year tour of duty at CHAD had ended in 1957.

"I did a full tour of duty here and only today do I realize just how big La Martinerie was," said Shilling, a member of the 110th Communications Squadron. He had traveled from his home in Shoreline, Washington, to take part in what was becoming a melancholy weekend. "It was immense. I know it's hard to believe, but during my three years at the base, I never saw these huge warehouses."

A week earlier, former Air Force brat Pam Kelley arrived in Châteauroux with a friend, determined to see where she was born. The forty-four year old Comcast Human Resources Director from Charleston, South Carolina had a premonition that she might not have another opportunity to get inside of what was once the 7373rd Air Force Hospital (AFH), now the French Army Headquarters.

"I believe I was one of the last babies born at the hospital," Kelley said. "In fact, there is a chance that I could have been the last baby born at the base. For years it has been my ambition to come here and actually see the hospital."

It was an ambition never to be realized. What happened to Pam underscored that the CHAD death knell was pervasive, and that even small common courtesies could be denied. Pam had never been to Châteauroux, but through her parents had learned of the Joe's From Maine legend and figured that the American-style eatery was a good place to find someone to help her in her quest. By amazing coincidence, Mike Gagné, son of the restaurant's namesake, was there to offer assistance. Gagné and I had been cleared for a two-man tour of the base the following day, and Pam agreed to follow us in her rented car.

Once on the base, we were shadowed by a one-car French military escort with a junior officer who had been instructed to never let us out of his sight. "I'm sorry, but I have my orders," the officer said, "I haven't been home in three months and I'm about to begin my leave in a few days. I can't afford to foul up and lose my chance to visit my family."

Mike, myself, and my wife, Darlene, dropped off Pam and her friend at a parking lot near the hospital while we wandered about, taking what would be the last photos of buildings that had played a prominent part in the base's history. A short time later, a distraught Pam approached us. "They turned me back," she said. "They wouldn't let me go inside, said it was restricted. I explained who I was, and explained that I only wanted to see where I was born. I was told that it was out of the question.

"Seeing the delivery room where I was born was especially important today because today is my birthday," Pam explained. "That's the reason I devoted part of my vacation in Germany to this trip, so I could have a real birthday celebration here at Châteauroux."

We drove Pam back to her car, hugged, offered condolences, and watched as she and her companion drove away.

This incident provided sharp contrast to another visit in Spring 2007 by the wife of an airman who had given birth to a daughter at the same hospital from which Pam was turned away. Colonel François LaPlace, then Commander of the 517th Régiment Du Train at La Martinerie, brought the incident to life.

"You know, we are in a room which was part of the former hospital. My office here is actually the room where women delivered babies. Sometimes I think of those people who now live on the other side of the ocean, and who were born in this very room fifty years ago.

"A few months ago, a woman asked me if she could visit the hospital where her daughter was born. Actually, it's my office, and she was so happy to come back to the place where her 'little one' was born. And it's funny to notice that this 'little one' is seven years older than me!" said LaPlace.

In 2007, LaPlace's transportation regiment seemed to be in perpetual motion, rotating units into heavily contested combat regions of Kosovo and Afghanistan. The regiment's mission was well defined as it fulfilled France's NATO commitment around the world. Enlisted men and officers alike knew and understood what was expected of them.

Three years later, the regiment was being disbanded and the stress brought on by anticipated action in remote regions of the world was accompanied by relocation jitters in France. Men and women would have different insignia on their uniform after they were split up and assigned to unfamiliar units throughout the country. As Pam Kelley found out, public relations cordiality can take a beating during times likes this.

A surprise awaited us at the Officers' Club, the fabled bastion of fun and games, imported fashion shows, Broadway revivals, good food, heavy drinking, a glass-targeted fireplace, slot machines, Saturday night striptease, and pretty women. Two earlier visits revealed a somber, padlocked building with weeds and overgrown plants crowding the front entrance and rear patio.

In early June 2010, the forbidding outside areas had been cleared, the front door was wide open, and you could see the chandeliers and mirrored walls that brightened the large foyer.

When we attempted to enter the building, we were turned away by our military shadow because the building was on the off limits list. We were saved by a woman who had just parked her car and identified herself as the President of the French Officers' Wives Club. She waved off our escort and took us inside. We discovered that the sin and decadence of a bygone era had been transformed into the women's Arts and Crafts Club, with the legendary bar area being the club's centerpiece. The large room was beautifully maintained, but there wasn't a bottle of booze, whiskey or beer glass in sight. The ghosts of the past were maintained by four beautifully handcrafted wooden beer taps extending like sentinels about twelve inches above the middle of the bar. It would be foolish to hazard a guess as to how many gallons of foamy brew owed their short existence to these instruments of pleasure.

It was evidently painting day for the women, and about a dozen of them clustered with their paint and brushes around large tables. Darlene, something of a self-proclaimed art critic, said most of them were not bad. We were largely ignored as we roamed about, took several photos, offered our thanks, and left. It became easy to understand why Barbara Bush, a Civil Service office worker, told me that the Officers' Club helped provide the best five years of her life.

For some unfathomable reason, our escort left us off at the main gate and said from that point we were on our own. It took no more than thirty seconds for us to be cleared through another gate with a single guard. Our main destination was the old high school, the memory of which had inspired an anonymous former student to donate a small fortune for a sculpture as a perpetual reminder of the Golden Ghetto's glory days. What we found were long neglected ruins that would have been better placed in an abandoned inner-city, low-rent housing project. One of the three, single floor high school buildings was missing. Gone—just gone—and nobody is quite sure what happened to it. Without an escort, we were free to roam in and about the two remaining buildings and poke around in what had once been the two-story junior high school only a few, weed-tangled yards away.

Eight days later, this freedom would be denied to the bus-load of pilgrims who were guided every step of their way by an affable Captain Daniel Fermon. The young officer spoke English with a charming French accent, was knowledgeable, smiled a lot, while making it clear that his charges went only where he allowed them and the schools were definitely off limits.

The outside of the buildings offered warning enough as to what would be found inside. One wooden front door was cracked and rotting, with rusty hinges barely able to hold it upright. The second front door leaned at an angle and was supported by a large, log-like timber. Almost all of the windows were broken in both buildings, steam radiators were pulled from the walls and laid on the floor of almost every classroom, large chunks of plaster had fallen from the walls, and almost all of the classroom doors had been pulled from their hinges. Wind-blown debris and rubbish had piled up in classroom corners, cracked floor tiles were everywhere, tangled and exposed electrical wiring had been pulled from long dead transformers and latrines were boarded up.

Much the same could be found at the old junior high school, which, surprisingly, still had a relatively secure stairway to the second floor. There was an eerie, flip-flapping of rotting window drapes in two of the upstairs rooms. Propelled by a slight breeze that wafted through the building, the drapes could be seen from the ground below, perfecting a silent message for the pilgrims who would arrive little more than a week later: "Why are you here? Keep your memories, cherish them— that was then and this is now."

One year after my depressing tour of the CHAD school grounds, I discovered enthusiasm in the most unlikely circumstance. I finally met Marguerite, the woman who had arrived in Châteauroux more than a half-century earlier and who, for me, had been more myth than reality. Our encounter was at a July 4th fete put together by the *France Etats-Unis Indre*, appropriately enough, at the former U.S. enclave of Brassioux. Now well into her eighties, Marguerite was impeccably dressed in summer white, with her red hair coiffed to perfection. Her jewelry was understated but obviously chosen with care. Marguerite's energy overwhelmed me, all smiles, warmth, penetrating eyes, the embodiment of the French expression, *bien dans sa peau*. It was hard to imagine that this vibrant lady had given birth to nine kids, each fathered by a different American GI, while working as a prostitute.

"I'm Marguerite and you, I think, are the American who has written a book about the base. You know, I think I deserve a book, too," she said with a big grin, and clinked her wine glass with mine.

Darlene had spoken to Marguerite earlier in the evening, remarking that their chemically enhanced hair colors were close to a perfect match. It was past the time for Darlene to get a touch up and *voilà*, Marguerite came to the rescue. Less than a week later, she guided my wife across *rue Diderot* in Châteauroux, grasping her arm as she led her into *La Haute Coiffure Française* not far from the Mayor's office. It was a classic celebrity tour de force. With Darlene in tow she greeted the dozen or so clients and stylists with a loud and resounding, *Bonjour touts*. There were familiar smiles all around as she introduced Darlene to her stylist, Pili, and fixed an appointment for the following week. Marguerite, with a full calendar for the day, turned down Darlene's offer of a coffee at the *Café de Paris*. Still holding her hand outside on the sidewalk, she said, "I hope your husband sells many, many books." She turned and walked away, a woman who had arrived in Châteauroux full of hope but with no job skills except one. Marguerite's strong stride was that of an elegant woman at ease with her past and basking in the local notoriety it had given her.

AMERICANS, WHAT AMERICANS?

During our annual Spring trip to France in 2010, we decided to take another look at the former hunting lodge of Ferdinand de Lesseps of Suez Canal fame in the village of Meunet-Planches. Our first visit two years earlier was in the early evening and darkness had begun to set in, hindering a really good look at the property. The sprawling, beautifully manicured estate was now listed as *Parc de Lesseps du Chateau de Planches* in Berry's *Parcs et Jardins* guide. Although not quite a tourist attraction, the large chateau and its connecting guest quarters had been drawing visitors who can discover a forest of some of the largest and most exotic trees in the region, each of them identified by a wooden plaque.

After parking our car, we saw something we had missed during our first visit despite its prominence and size. Centered in a lawn off to the side of a parking area is a large, Neoclassical sculpture of a bare-chested laborer, a depiction of the heroic endeavor necessary to construct the Suez Canal.

We saw lights in the window of one of the chateau's wings. An expensive late-model car was parked not far away. It was mid-Sunday afternoon and the estate appeared to be deserted. Our knock on the door was answered by a middle-aged man wearing a sports jacket emblazoned with the logo of a Las Vegas casino. He called inside and a proprietor, a well-dressed woman "of a certain age" emerged. She was stylish in slacks and a coordinating knit top with discrete gold jewelry. She was somewhat distant, could hardly be called outgoing, and who could blame her? We were interrupting her Sunday afternoon at home. Nonetheless, she gave us permission to walk around, take pictures, and tour a portion of the surrounding grounds.

When we returned, we found that the door had been closed and the car had been moved into a position in front of the entrance. Despite this, I decided that it was appropriate to thank her. She answered my knock and I felt that it was a good time to explain to her the reason for our visit. Much to my surprise, she listened with interest and a thin smile enlivened her face. I told her I had written a book about the Châteauroux Air Station, a sizeable part of which was devoted to the de Lesseps hunting lodge. Her smile broadened when I told her that for thirteen to fourteen years during the 1950s and 1960s, high-ranking American civilians and later Air Force officers rented the chateau. Wild parties and ribald behavior were the norm.

Her reply reaffirmed a theme that had threaded its way throughout *Golden Ghetto*'s narrative, simply that it is a flimsy fabric that holds memories together, attempting despite often great odds to keep intact the history they convey.

"My family purchased this place in 1972 and we had been told many times that Americans had occupied the chateau," she said. "However, this is the very first time that it has been confirmed that Americans had ever really lived here at all."

The French Inherit Shangri-La

The friendly and quizzical encounter at Meunet-Planches occurred one year before a cornucopia crammed with Franco-American history was emptied for all to see. There might not have been specific mention of the high level cavorting and merry-making at the de Lesseps hunting lodge, but there were riches galore for even the greediest inquisitor into the American experience in Berry.

On July 9, 2011, a three-month exhibition like none other before opened at *la mediatheque de Châteauroux: Châteauroux 1951–1967, c'etait l'Amerique!* (Châteauroux 1951 – 1967, it was America!). The exhibit took over an entire wing just off the main entrance of the city library. It was the culmination of a project that originated among the instructors and gifted school kids at Châteauroux's private *lycée*, Sainte-Solange. Support for the project seems to have come from everywhere; archivists, a writers task force headed by Jean-Paul Jody, the tireless, self-taught historian Yves Bardet, an organization of former employees at the American base (AECBA), documentarian Anice Clement and French historian Pierre Remerand. It was their collective job to validate the texts produced by the students and to ensure that the memorabilia on exhibit was an accurate portrayal of the base's history.

The Saturday began with threatening skies and occasional light rain. This did not temper the expectations of a large cross section of Berrichons who had received invitations to the inaugural reception. It was typically French. What is a reception without good sparkling wine and tasty hors-d'oeuvres? There was plenty of each, despite the freeloading interlopers who muscled in from the book stacks. Before the treats were offered there were, of course, the speeches—many of them. Most elicited signs of relief because of their brevity but there was no avoiding the tedium that accompanied the longer VIP screeds. Darlene and I were surprised by how smoothly everything came together.

Reunions in 1998, 2007 and the 2009 unveiling of The Flame of Friendship statue were extraordinary demonstrations of the power that rekindled, happy memories and friendships can have. It made little difference how long the returning pilgrims had been at CHAD—whether one, two, three years or longer—it was the best time of their lives. Everyone was young, the men were virile, the women sexy, opportunities for youthful debauchery were everywhere, marriages were made and you could even get American-style French fries, hamburgers, chili, and hot dogs at Joe's From Maine.

As I wandered through the library exhibit that Saturday morning, it became clear to me that the CHAD talisman had been passed onto a new generation, not American but French. Unlike the reunions of American pilgrims, the curious French supplicants were petitioning the past to understand the present and hopefully shape their futures. The aged American pilgrims without realizing it had, at the unveiling of The Flame of Friendship a year earlier, effectively passed on the keys to Shangri-La. As the years dwindle, the likelihood of future American reunions has become improbable if not impossible.

The library reception by Châteauroux standards took on epic proportion as it progressed. The exhibit area became increasingly crowded as the word got out. It also became a multigenerational affair—grandmothers and grandfathers with their sons and daughters and grandchildren moved from one display to another. "I was told that you are an American," a sedate woman with horn-rimmed glasses and red dress in her thirties approached me inquiringly, her English in good order. "I am a teacher at the *lycée* Sainte-Solange, my students have participated in this project from its start. I was told that you have written a book about the base, is that true? Were you at the base when it was open? Will you still be here in September?" She didn't introduce herself by name and I forgot to ask.

This little lady was a repository for direct questions. Unfortunately, my responses to two of her questions disappointed her. No, I told her, I was never stationed at the base. And no, I would not be around in September. "That is too bad, my students would have wanted to speak with you, a real American."

It seemed like everything and anything that could be uncovered that dealt with CHAD was on display. Over here was a slot machine from one of the clubs on the base, simplistic by today's standards but a top performer fifty years earlier. Next to the entry to the media viewing room was a pinball machine with the voluptuous image of Marilyn Monroe as its crowning glory. On display on top of an American flag was a jukebox push-button remote unit, a ubiquitous necessity in booths in every American diner, soda and ice cream parlor, and undoubtedly at the snack bar at the base. Yves Bardet's matchless schematic rendering of Brassioux was given the importance it deserved.

I lingered for a while near the glass case of newspaper accounts of the JFK assassination. Visitors would stop momentarily, acknowledge with nods, and quickly move on. This was a reception devoted to good memories.

Only steps away, another case displayed the fun stuff—literature for kids old and young, with comic books as the central focus. Appropriately, one comic book chronicled the questionable exploits of Sergeant Bilco, an Army master sergeant who seemed to have nothing more to do than hatch get-rich-quick schemes and devise ways to avoid work. Scrutinizing it all was a life-size mannequin in full dress Air Force uniform complete with sergeant stripes and ribbons.

It would take a week for even the fastest speed reader to absorb all of the newspaper accounts that festooned the walls and crammed the glass cases. Every nuance was explored, with visitor reactions described the next day by *La Nouvelle République* in a very straightforward manner, "1967: People Bemoan the American Departure." Forty-four years after their departure, a bunch of talented Châteauroux school kids began their intellectual journey to discover the reasons why. Their youthful curiosity would inevitably take them through the gate to Shangri-La, where they would uncover the true meaning of *The Golden Ghetto.*

BIOGRAPHICAL NOTE

Steve Bassett was born, raised, and educated in New Jersey before joining the dwindling number of itinerant newsmen roaming the countryside in search of just about everything. Stints as a featured reporter with newspapers in New Jersey, Illinois, and Salt Lake City were followed by Associated Press assignments in Phoenix and finally as Special Urban Affairs writer in San Francisco. Then came CBS television news, three Emmy Awards for his investigative documentaries, and the prestigious first Medallion Award, presented by the California Bar Association for "Distinguished Reporting on the Administration of Justice." Along the way he found time to author *The Battered Rich* (Ashley Books), exposing seldom discussed but widespread marital abuse among the affluent. He lives in Placitas, New Mexico.